# Advanced Topics in Levitation

# Contents

# Chapter 1

# Levitation

This article is about the scientific techniques of levitation. For the magic illusion, see Magic (illusion). For paranormal activity, see Levitation (paranormal). For other uses, see Levitation (disambiguation).

**Levitation** (from Latin *levitas* "lightness")[1] is the process by which an object is held aloft, without mechanical support,

*A cubical magnet levitating over a superconducting material (known as the Meissner effect)*

in a stable position.

Levitation is accomplished by providing an upward force that counteracts the pull of gravity (in relation to gravity on earth), plus a smaller stabilizing force that pushes the object toward a home position whenever it is a small distance away from that home position. The force can be a fundamental force such as magnetic or electrostatic, or it can be a reactive force such as optical, buoyant, aerodynamic, or hydrodynamic.

Levitation excludes floating at the surface of a liquid because the liquid provides direct mechanical support. Levitation excludes hovering flight by insects, hummingbirds, helicopters, rockets, and balloons because the object provides its own counter-gravity force.

## 1.1   Physics

Levitation (on Earth or any planetoid) requires an upward force that cancels out the weight of the object, so that the object does not fall (accelerate downward) or rise (accelerate upward). For positional stability, any small displacement of the levitating object must result in a small change in force in the opposite direction. The small changes in force can be accomplished by gradient field(s) or by active regulation. If the object is disturbed, it might oscillate around its final position, but its motion eventually decreases to zero due to damping effects. (In a turbulent flow, the object might oscillate indefinitely.)

Levitation techniques are useful tools in physics research. For example, levitation methods are useful for high-temperature melt property studies because they eliminate the problem of reaction with containers and allow deep undercooling of melts. The containerless conditions may be obtained by opposing gravity with a levitation force instead of allowing an entire experiment to freefall.[2]

## 1.2   Magnetic levitation

Main article: Magnetic levitation
Magnetic levitation is the most commonly seen and used form of levitation.

Diamagnetic materials are commonly used for demonstration purposes. In this case the returning force appears from the interaction with the screening currents. For example, a superconducting sample, which can be considered either as a perfect diamagnet or an ideally hard superconductor, easily levitates in an ambient external magnetic field. The superconductor is first heated strongly, then cooled with liquid nitrogen to levitate on top of a diamagnet. In a very strong magnetic field by means of diamagnetic levitation, even small live animals have been levitated.

It is possible to levitate pyrolytic graphite by placing thin squares of it above four cube magnets with the north poles forming one diagonal and south poles forming the other diagonal.[3]

Magnetic levitation is in development for use for transportation systems. For example, the Maglev includes trains that are levitated by a large number of magnets. Due to the lack of friction on the guide rails, they are faster, quieter, and smoother than wheeled mass transit systems.

Electrodynamic suspension uses AC magnetic fields.

## 1.3   Electrostatic levitation

Main article: Electrostatic levitation

In electrostatic levitation an electric field is used to counteract gravitational force.

## 1.4   Aerodynamic levitation

Main article: Aerodynamic levitation

*a high-temperature superconductor levitating above magnet*

*A magnetically levitated (maglev) train departing Pudong International Airport on the first commercial high-speed maglev line in the world.*

In aerodynamic levitation, the levitation is achieved by floating the object on a stream of gas, either produced by the object or acting on the object. For example, a ping pong ball can be levitated with the stream of air from a vacuum cleaner set on 'blow'. With enough thrust, very large objects can be levitated using this method.

### 1.4.1  Gas film levitation

This technique enables the levitation of an object against gravitational force by floating it on a thin gas film formed by gas flow through a porous membrane. Using this technique, high temperature melts can be kept clean from contamination and be supercooled.[2] A common example in general usage includes air hockey, where the puck is lifted by a thin layer of air. Hovercraft also use this technique, producing a large region of high-pressure air underneath them.

## 1.5  Acoustic levitation

Main article: Acoustic levitation

Acoustic levitation uses sound waves to provide a levitating force.

## 1.6  Optical levitation

Main article: Optical levitation

Optical levitation is a technique in which a material is levitated against the downward force of gravity by an upward force stemming from photon momentum transfer (radiation pressure).

## 1.7  Buoyant levitation

Gasses at high pressure can have a density exceeding that of some solids. Thus they can be used to levitate solid objects through buoyancy.[4] Noble gases are preferred for their non-reactivity. Xenon is the densest non-radioactive noble gas, at 5.894g/L. Xenon has been used to levitate polyethylene, at a pressure of 154atm.

## 1.8  Casimir force

Scientists have discovered a way of levitating ultra small objects by manipulating the so-called Casimir force, which normally causes objects to stick together due to forces predicted by quantum field theory. This is, however, only possible for micro-objects.[5][6]

## 1.9  Uses

### 1.9.1  Maglev trains

Main article: Maglev

Magnetic levitation is used to suspend trains without touching the track. This permits very high speeds, and greatly reduces the maintenance requirements for tracks and vehicles, as little wear occurs. This also means there is no friction, so the only force acting against it is air resistance.

### 1.9.2  Animal levitation

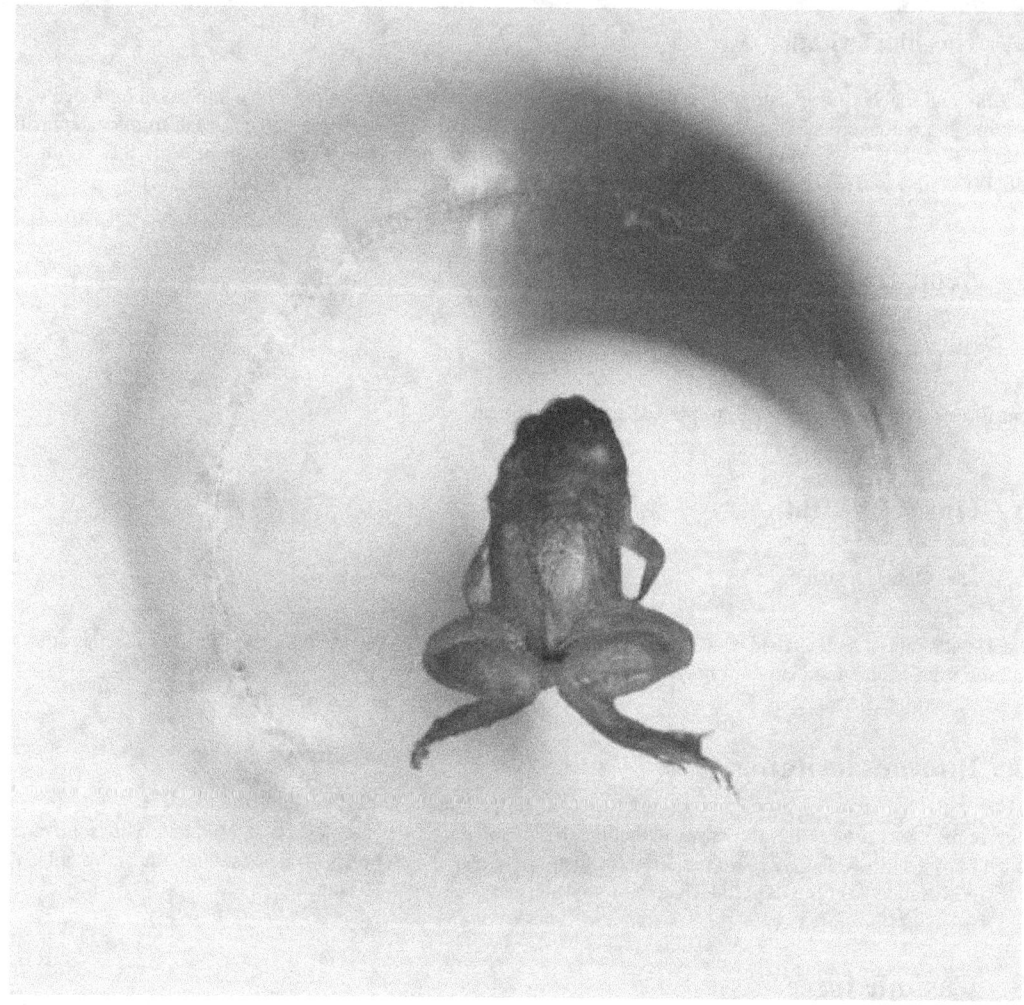

*Diamagnetic levitation of a live frog.*

Scientists have levitated frogs,[7] grasshoppers, and mice by means of powerful electromagnets utilizing superconductors, producing diamagnetic repulsion of body water. The mice acted confused at first, but adapted to the levitation after approximately four hours, suffering no immediate ill effects.[8][9]

## 1.10   Further reading

- Charles P. Strehlow, M. C. Sullivan (2008). "A Classroom Demonstration of Levitation...". arXiv:0803.3090..

## 1.11   See also

- Anti-gravity

- Flight

- Weightlessness

## 1.12   References

[1] *Levitate*, "to rise by virtue of lightness," from Latin *levitas* "lightness," patterned in English on *gravitate*: Online Etymology Dictionary

[2] Paul C. Nordine, J. K. Richard Weber, and Johan G. Abadie (2000), "Properties of high-temperature melts using levitation", *Pure and Applied Chemistry* **72**: 2127–2136, doi:10.1351/pac200072112127

[3] Waldron, Robert D., "Diamagnetic Levitation Using Pyrolytic Graphite", *Review of Scientific Instruments* **37**: 29–35, Bibcode:1966RScI...37...29W, doi:10.1063/1.1719946

[4] http://www.mrs.org/s_mrs/sec_subscribe.asp?CID=12048&DID=275340&action=detail Materials Processing Through Levitation in High Gas Pressure

[5] Scientists reveal secret of levitation, Yahoo! News

[6] Levitation in Miniature, Null Hypothesis

[7] "Frogs Levitate in a strong enough magnetic field". physics.org. Retrieved 20 November 2014.

[8] "NASA Levitates a Mouse With Magnetic Fields". *Popular Science*. September 9, 2009. Retrieved 20 November 2014.

[9] Mice Levitated in Lab

## 1.13   External links

The dictionary definition of levitation at Wiktionary

- Diamagnetic Levitation (YouTube)

- Superconducting Levitation Demos

# Chapter 2

# Acoustic levitation

**Acoustic levitation** (also: **Acoustophoresis**) is a method for suspending matter in a medium by using acoustic radiation pressure from intense sound waves in the medium.

Sometimes ultrasonic frequencies can be used to levitate objects, thus creating no sound heard by the human ear, such as was demonstrated at Otsuka Lab,[1] while others use audible frequencies. There are various ways of emitting the sound wave, from creating a wave underneath the object and reflecting it back to its source, to using a (transparent) tank to create a large acoustic field.

Acoustic levitation is usually used for containerless processing which has become more important of late due to the small size and resistance of microchips and other such things in industry. Containerless processing may also be used for applications requiring very-high-purity materials or chemical reactions too rigorous to happen in a container. This method is harder to control than other methods of containerless processing such as electromagnetic levitation but has the advantage of being able to levitate nonconducting materials.

By 2013, acoustic levitation had progressed from motionless levitation to controllably moving hovering objects, an ability useful in the pharmaceutical and electronics industries.[2] A prototype device involved a chessboard-like array of square acoustic emitters that move an object from one square to another by slowly lowering the sound intensity emitted from one square while increasing the sound intensity from the other.[2]

Current systems have lifted at most a few kilograms.[3] Acoustic levitators are used mostly in industry however some products are commercially available to the public.

## 2.1 See also

- Acoustic tweezers
- Optical levitation
- Radiation pressure
- Electrostatic levitation
- Magnetic levitation
- Aerodynamic levitation
- Buoyancy

## 2.2 References

[1] "Ultrasonic Levitation". Archived from the original on 4 November 2006. Retrieved 2006-11-15.

[2] Kim, Meeri (July 15, 2013). "Sound waves can be used to levitate and move objects, study says". *The Washington Post*. Archived from the original on July 16, 2013. Kim's article cites Foresti, Daniele et al. (July 10, 2013--date based on URL). "Acoustophoretic contactless transport and handling of matter in air (Abstract)". *Proceedings of the National Academy of Sciences*. Archived from the original on July 16, 2013. Check date values in: |date= (help)

[3] "Phenomena, theory and applications of near-field acoustic levitation" (PDF). Archived from the original (PDF) on 2005-01-16.

## 2.3  External links

- McGraw-Hill AccessScience: Acoustic radiation pressure

- A Multi-Transducer Near Field Acoustic Levitation System for Noncontact Transportation of Large-Sized Planar Objects

- Live Science – Scientists Levitate Small Animals

# Chapter 3

# Aerodynamic levitation

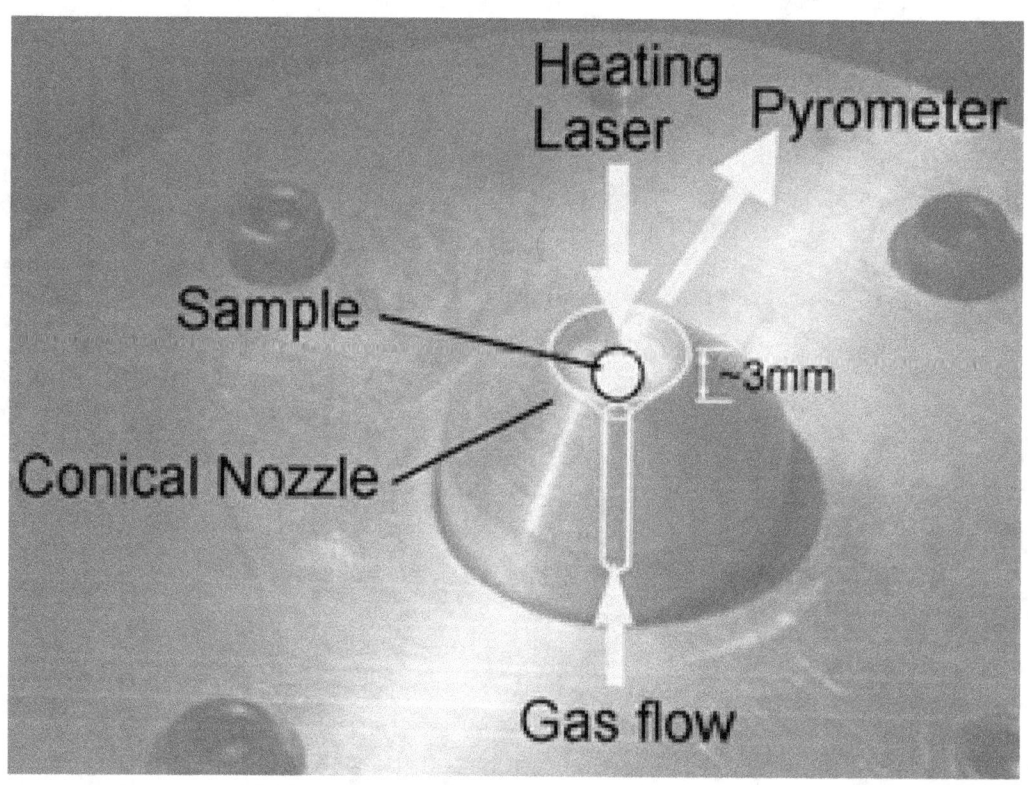

*Aerodynamic levitation apparatus: a spherical sample is floated on a gas stream which flows through the conical nozzle. Sample is heated by a CO$_2$ laser and temperature is measured from sample brightness by a pyrometer.*

**Aerodynamic levitation** is the use of gas pressure to levitate materials so that they are no longer in physical contact with any container. In scientific experiments this removes contamination and nucleation issues associated with physical contact with a container.

9

## 3.1   Overview

The term aerodynamic levitation could be applied to many objects that use gas pressure to counter the force of gravity, and allow stable levitation. Helicopters and air hockey pucks are two good examples of objects that are aerodynamically levitated. However, more recently this term has also been associated with a scientific technique which uses a cone-shaped nozzle allowing stable levitation of 1-3mm diameter spherical samples without the need for active control mechanisms.[1]

## 3.2   Aerodynamic levitation as a scientific tool

These systems allow spherical samples to be levitated by passing gas up through a diverging conical nozzle. Combining this with >200W continuous $CO_2$ laser heating allows sample temperatures in excess of 3000 degrees Celsius to be achieved.

When heating materials to these extremely high temperatures levitation in general provides two key advantages over traditional furnaces. First, contamination that would otherwise occur from a solid container is eliminated. Second, the sample can be undercooled, i.e. cooled below its normal freezing temperature without actually freezing.

### 3.2.1   Undercooling of liquid samples

Undercooling, or supercooling, is the cooling of a liquid below its equilibrium freezing temperature while it remains a liquid. This can occur wherever crystal nucleation is suppressed. In levitated samples, heterogeneous nucleation is suppressed due to lack of contact with a solid surface. Levitation techniques typically allow samples to be cooled several hundred degrees Celsius below their equilibrium freezing temperatures.

### 3.2.2   Glass produced by aerodynamic levitation

Since crystal nucleation is suppressed by levitation, and since it is not limited by sample conductivity (unlike electromagnetic levitation), aerodynamic levitation can be used to make glassy materials, from high temperature melts that cannot be made by standard methods. Several silica-free, aluminium oxide based glasses have been made.[2][3][4]

### 3.2.3   Physical property measurements

In the last few years a range of in situ measurement techniques have also been developed. The following measurements can be made with varying precision:

electrical conductivity, viscosity, density, surface tension, specific heat capacity,

In situ aerodynamic levitation has also been combined with:

X-ray synchrotron radiation, neutron scattering, NMR spectroscopy

## 3.3   See also

- Magnetic levitation

- Electrostatic levitation

- Optical levitation

- Acoustic levitation

## 3.4   Further reading

- Price, D.L. (2010). *High-Temperature Levitated Materials*. Cambridge University Press. ISBN 0521880521.

## 3.5   References

[1] Paul C. Nordine; J. K. Richard Weber & Johan G. Abadie (2000), "Properties of high-temperature melts using levitation", *Pure and Applied Chemistry* **72** (11): 2127–2136, doi:10.1351/pac200072112127

[2] J. K. Richard Weber; Jean A. Tangeman; Thomas S. Key; Kirsten J. Hiera; Paul-Francois Paradis; Takehiko Ishikawa; et al. (2002), "Novel Synthesis of Calcium Oxide–Aluminum Oxide Glasses", *Japanese Journal of Applied Physics* **41**: 3029–3030, Bibcode:2002JaJAP..41.3029W, doi:10.1143/JJAP.41.3029

[3] J. K. Richard Weber; Johan G. Abadie; April D. Hixson; Paul C. Nordine; Gregory A. Jerman (2004), "Glass Formation and Polyamorphism in Rare-Earth Oxide–Aluminum Oxide Compositions", *Journal of the American Ceramic Society* **83** (8): 1868–1872, doi:10.1111/j.1151-2916.2000.tb01483.x

[4] L. B. Skinner; A. C. Barnes & W. Crichton (2006), "Novel behaviour and structure of new glasses of the type Ba–Al–O and Ba–Al–Ti–O produced by aerodynamic levitation and laser heating", *Journal of Physics: Condensed Matter* **18** (32): L407–L414, Bibcode:2006JPCM...18L.407S, doi:10.1088/0953-8984/18/32/L01

# Chapter 4

# Casimir effect

In quantum field theory, the **Casimir effect** and the **Casimir–Polder force** are physical forces arising from a quantized field. They are named after the Dutch physicist Hendrik Casimir.

The typical example is of two uncharged conductive plates in a vacuum, placed a few nanometers apart. In a classical description, the lack of an external field means that there is no field between the plates, and no force would be measured between them.[1] When this field is instead studied using the QED vacuum of quantum electrodynamics, it is seen that the plates do affect the virtual photons which constitute the field, and generate a net force[2]—either an attraction or a repulsion depending on the specific arrangement of the two plates. Although the Casimir effect can be expressed in terms of virtual particles interacting with the objects, it is best described and more easily calculated in terms of the zero-point energy of a quantized field in the intervening space between the objects. This force has been measured and is a striking example of an effect captured formally by second quantization.[3][4] However, the treatment of boundary conditions in these calculations has led to some controversy. In fact, "Casimir's original goal was to compute the van der Waals force between polarizable molecules" of the conductive plates. Thus it can be interpreted without any reference to the zero-point energy (vacuum energy) of quantum fields.[5]

Dutch physicists Hendrik Casimir and Dirk Polder at Philips Research Labs proposed the existence of a force between two polarizable atoms and between such an atom and a conducting plate in 1947, and, after a conversation with Niels Bohr who suggested it had something to do with zero-point energy, Casimir alone formulated the theory predicting a force between neutral conducting plates in 1948; the former is called the Casimir–Polder force while the latter is the Casimir effect in the narrow sense. Predictions of the force were later extended to finite-conductivity metals and dielectrics by Lifshitz and his students, and recent calculations have considered more general geometries. It was not until 1997, however, that a direct experiment, by S. Lamoreaux, described above, quantitatively measured the force (to within 15% of the value predicted by the theory),[6] although previous work [e.g. van Blockland and Overbeek (1978)] had observed the force qualitatively, and indirect validation of the predicted Casimir energy had been made by measuring the thickness of liquid helium films by Sabisky and Anderson in 1972. Subsequent experiments approach an accuracy of a few percent.

Because the strength of the force falls off rapidly with distance, it is measurable only when the distance between the objects is extremely small. On a submicron scale, this force becomes so strong that it becomes the dominant force between uncharged conductors. In fact, at separations of 10 nm—about 100 times the typical size of an atom—the Casimir effect produces the equivalent of about 1 atmosphere of pressure (the precise value depending on surface geometry and other factors).[7]

In modern theoretical physics, the Casimir effect plays an important role in the chiral bag model of the nucleon; in applied physics, it is significant in some aspects of emerging microtechnologies and nanotechnologies.[8]

Any medium supporting oscillations has an analogue of the Casimir effect. For example, beads on a string[9][10] as well as plates submerged in noisy water[11] or gas[12] illustrate the Casimir force.

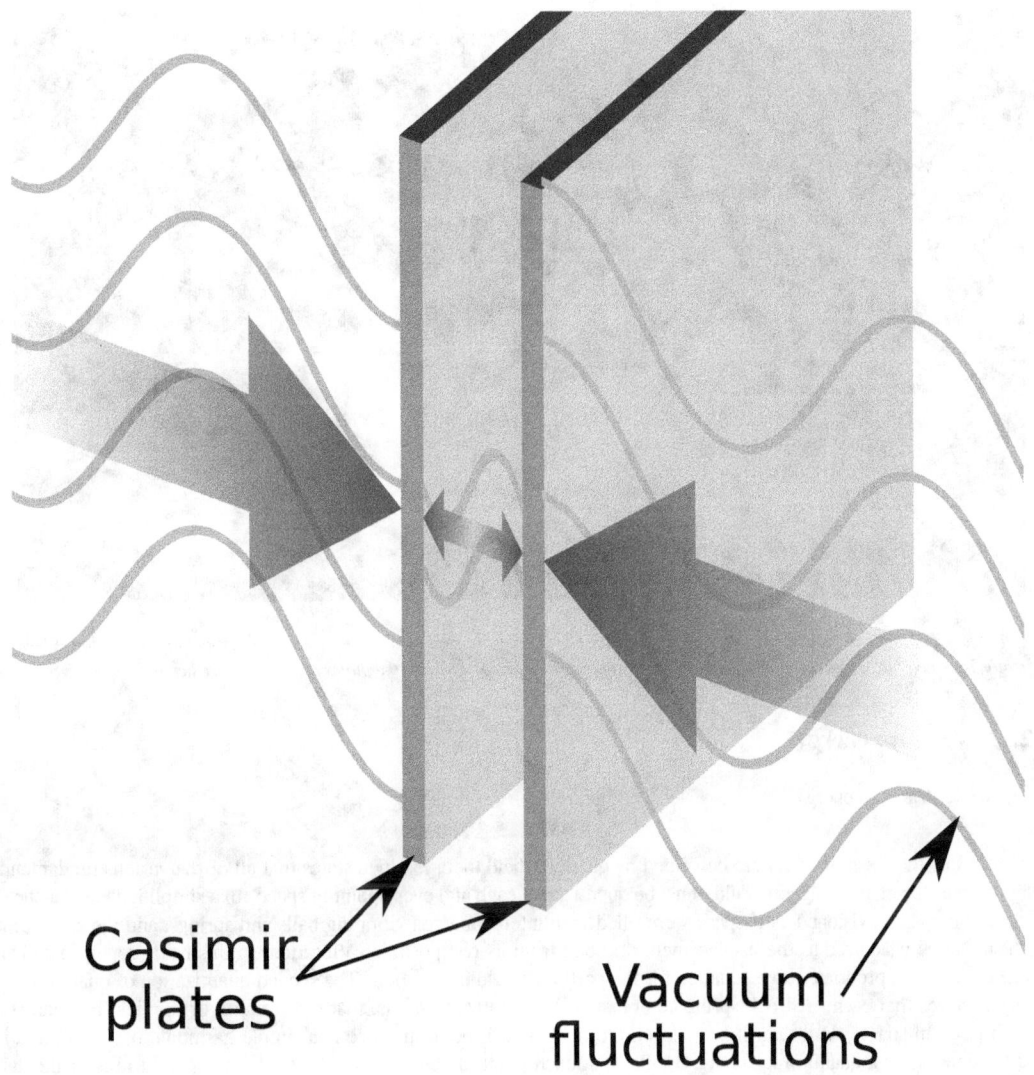

*Casimir forces on parallel plates*

## 4.1 Overview

The Casimir effect can be understood by the idea that the presence of conducting metals and dielectrics alters the vacuum expectation value of the energy of the second quantized electromagnetic field.[13][14] Since the value of this energy depends on the shapes and positions of the conductors and dielectrics, the Casimir effect manifests itself as a force between such objects.

## 4.2 Possible causes

*A water wave analogue of the Casimir effect. Two parallel plates are submerged into colored water contained in a sonicator. When the sonicator is turned on, waves are excited imitating vacuum fluctuations; as a result, the plates are forced together.*

### 4.2.1 Vacuum energy

Main article: Vacuum energy

The causes of the Casimir effect are described by quantum field theory, which states that all of the various fundamental fields, such as the electromagnetic field, must be quantized at each and every point in space. In a simplified view, a "field" in physics may be envisioned as if space were filled with interconnected vibrating balls and springs, and the strength of the field can be visualized as the displacement of a ball from its rest position. Vibrations in this field propagate and are governed by the appropriate wave equation for the particular field in question. The second quantization of quantum field theory requires that each such ball-spring combination be quantized, that is, that the strength of the field be quantized at each point in space. At the most basic level, the field at each point in space is a simple harmonic oscillator, and its quantization places a quantum harmonic oscillator at each point. Excitations of the field correspond to the elementary particles of particle physics. However, even the vacuum has a vastly complex structure, so all calculations of quantum field theory must be made in relation to this model of the vacuum.

The vacuum has, implicitly, all of the properties that a particle may have: spin, or polarization in the case of light, energy, and so on. On average, most of these properties cancel out: the vacuum is, after all, "empty" in this sense. One important exception is the vacuum energy or the vacuum expectation value of the energy. The quantization of a simple harmonic oscillator states that the lowest possible energy or zero-point energy that such an oscillator may have is

$$E = \tfrac{1}{2} \hbar \omega \ .$$

Summing over all possible oscillators at all points in space gives an infinite quantity. Since only *differences* in energy are physically measurable (with the notable exception of gravitation, which remains beyond the scope of quantum field theory), this infinity may be considered a feature of the mathematics rather than of the physics. This argument is the underpinning of the theory of renormalization. Dealing with infinite quantities in this way was a cause of widespread unease among quantum field theorists before the development in the 1970s of the renormalization group, a mathematical formalism for scale transformations that provides a natural basis for the process.

When the scope of the physics is widened to include gravity, the interpretation of this formally infinite quantity remains problematic. There is currently no compelling explanation as to why it should not result in a cosmological constant that is many orders of magnitude larger than observed.[15] However, since we do not yet have any fully coherent quantum theory of gravity, there is likewise no compelling reason as to why it should.[16]

### 4.2.2 Relativistic van der Waals force

Alternatively, a 2005 paper by Robert Jaffe of MIT states that "Casimir effects can be formulated and Casimir forces can be computed without reference to zero-point energies. They are relativistic, quantum forces between charges and currents. The Casimir force (per unit area) between parallel plates vanishes as alpha, the fine structure constant, goes to zero, and the standard result, which appears to be independent of alpha, corresponds to the alpha approaching infinity limit," and that "The Casimir force is simply the (relativistic, retarded) van der Waals force between the metal plates."[17]

### 4.2.3 Coupled ground-state energy

Finally, a third way of understanding Casimir forces has been suggested, based on canonical macroscopic quantum electrodynamics. In this interpretation, there exists a ground (vacuum) state of the *coupled* system of matter and fields, which determines the ground-state properties of the electromagnetic field, giving rise to a force. The Casimir force is fundamentally a property of the coupled system of matter and fields, in which the interaction between the plates is mediated by the zero-point fields. In more traditional interpretations, however, the emphasis has fallen either on the electromagnetic field or the fluctuating material in the plates.[18]

## 4.3 Effects

Casimir's observation was that the second-quantized quantum electromagnetic field, in the presence of bulk bodies such as metals or dielectrics, must obey the same boundary conditions that the classical electromagnetic field must obey. In particular, this affects the calculation of the vacuum energy in the presence of a conductor or dielectric.

Consider, for example, the calculation of the vacuum expectation value of the electromagnetic field inside a metal cavity, such as, for example, a radar cavity or a microwave waveguide. In this case, the correct way to find the zero-point energy of the field is to sum the energies of the standing waves of the cavity. To each and every possible standing wave corresponds an energy; say the energy of the $n$th standing wave is $E_n$. The vacuum expectation value of the energy of the electromagnetic field in the cavity is then

$$\langle E \rangle = \frac{1}{2} \sum_n E_n$$

with the sum running over all possible values of $n$ enumerating the standing waves. The factor of 1/2 corresponds to the fact that the zero-point energies are being summed (it is the same 1/2 as appears in the equation $E = \hbar\omega/2$). Written in this way, this sum is clearly divergent; however, it can be used to create finite expressions.

In particular, one may ask how the zero-point energy depends on the shape $s$ of the cavity. Each energy level $E_n$ depends on the shape, and so one should write $E_n(s)$ for the energy level, and $\langle E(s) \rangle$ for the vacuum expectation value. At this point comes an important observation: the force at point $p$ on the wall of the cavity is equal to the change in the vacuum energy if the shape $s$ of the wall is perturbed a little bit, say by $\delta s$, at point $p$. That is, one has

$$F(p) = -\left. \frac{\delta \langle E(s) \rangle}{\delta s} \right|_p .$$

This value is finite in many practical calculations.[19]

Attraction between the plates can be easily understood by focusing on the one-dimensional situation. Suppose that a moveable conductive plate is positioned at a short distance $a$ from one of two widely separated plates (distance $L$ apart). With $a \ll L$, the states within the slot of width $a$ are highly constrained so that the energy $E$ of any one mode is widely separated from that of the next. This is not the case in open region $L$, where there is a large number (about $L/a$) of states with energy evenly spaced between $E$ and the next mode in the narrow slot---in other words, all slightly larger than $E$. Now on shortening $a$ by d$a$ ($< 0$), the mode in the slot shrinks in wavelength and therefore increases in energy proportional to -d$a/a$, whereas all the outside $L/a$ states lengthen and correspondingly lower energy proportional to d$a/L$ (note the denominator). The net change is slightly negative, because all the $L/a$ modes' energies are slightly larger than the single mode in the slot.

## 4.4 Derivation of Casimir effect assuming zeta-regularization

- *See **Wikiversity** for an elementary calculation in one dimension.*

In the original calculation done by Casimir, he considered the space between a pair of conducting metal plates at distance $a$ apart. In this case, the standing waves are particularly easy to calculate, because the transverse component of the electric field and the normal component of the magnetic field must vanish on the surface of a conductor. Assuming the parallel plates lie in the xy-plane, the standing waves are

$$\psi_n(x, y, z; t) = e^{-i\omega_n t} e^{ik_x x + ik_y y} \sin(k_n z)$$

where $\psi$ stands for the electric component of the electromagnetic field, and, for brevity, the polarization and the magnetic components are ignored here. Here, $k_x$ and $k_y$ are the wave vectors in directions parallel to the plates, and

$$k_n = \frac{n\pi}{a}$$

is the wave-vector perpendicular to the plates. Here, $n$ is an integer, resulting from the requirement that $\psi$ vanish on the metal plates. The frequency of this wave is

$$\omega_n = c\sqrt{k_x{}^2 + k_y{}^2 + \frac{n^2\pi^2}{a^2}}$$

where $c$ is the speed of light. The vacuum energy is then the sum over all possible excitation modes. Since the area of the plates is large, we may sum by integrating over two of the dimensions in k-space. The assumption of periodic boundary conditions yields,

$$\langle E \rangle = \frac{\hbar}{2} \cdot 2 \int \frac{A dk_x dk_y}{(2\pi)^2} \sum_{n=1}^{\infty} \omega_n$$

where $A$ is the area of the metal plates, and a factor of 2 is introduced for the two possible polarizations of the wave. This expression is clearly infinite, and to proceed with the calculation, it is convenient to introduce a regulator (discussed in greater detail below). The regulator will serve to make the expression finite, and in the end will be removed. The zeta-regulated version of the energy per unit-area of the plate is

$$\frac{\langle E(s) \rangle}{A} = \hbar \int \frac{dk_x dk_y}{(2\pi)^2} \sum_{n=1}^{\infty} \omega_n |\omega_n|^{-s}.$$

In the end, the limit $s \to 0$ is to be taken. Here $s$ is just a complex number, not to be confused with the shape discussed previously. This integral/sum is finite for $s$ real and larger than 3. The sum has a pole at $s = 3$, but may be analytically continued to $s = 0$, where the expression is finite. The above expression simplifies to:

$$\frac{\langle E(s) \rangle}{A} = \frac{\hbar c^{1-s}}{4\pi^2} \sum_n \int_0^\infty 2\pi q dq \left| q^2 + \frac{\pi^2 n^2}{a^2} \right|^{(1-s)/2},$$

where polar coordinates $q^2 = k_x^2 + k_y^2$ were introduced to turn the double integral into a single integral. The $q$ in front is the Jacobian, and the $2\pi$ comes from the angular integration. The integral converges if Re[$s$] > 3, resulting in

$$\frac{\langle E(s) \rangle}{A} = -\frac{\hbar c^{1-s} \pi^{2-s}}{2a^{3-s}} \frac{1}{3-s} \sum_n |n|^{3-s}.$$

The sum diverges at $s$ in the neighborhood of zero, but if the damping of large-frequency excitations corresponding to analytic continuation of the Riemann zeta function to $s = 0$ is assumed to make sense physically in some way, then one has

$$\frac{\langle E \rangle}{A} = \lim_{s \to 0} \frac{\langle E(s) \rangle}{A} = -\frac{\hbar c \pi^2}{6a^3} \zeta(-3).$$

But

$$\zeta(-3) = \frac{1}{120}$$

and so one obtains

$$\frac{\langle E \rangle}{A} = \frac{-\hbar c \pi^2}{3 \cdot 240 a^3}.$$

The analytic continuation has evidently lost an additive positive infinity, somehow exactly accounting for the zero-point energy (not included above) outside the slot between the plates, but which changes upon plate movement within a closed system. The Casimir force per unit area $F_c / A$ for idealized, perfectly conducting plates with vacuum between them is

$$\frac{F_c}{A} = -\frac{d}{da} \frac{\langle E \rangle}{A} = -\frac{\hbar c \pi^2}{240 a^4}$$

where

$\hbar$ (hbar, ℏ) is the reduced Planck constant,

$c$ is the speed of light,

$a$ is the distance between the two plates

The force is negative, indicating that the force is attractive: by moving the two plates closer together, the energy is lowered. The presence of $\hbar$ shows that the Casimir force per unit area $F_c / A$ is very small, and that furthermore, the force is inherently of quantum-mechanical origin.

NOTE: In Casimir's original derivation , a moveable conductive plate is positioned at a short distance $a$ from one of two widely separated plates (distance $L$ apart). The 0-point energy on *both* sides of the plate is considered. Instead of the above *ad hoc* analytic continuation assumption, non-convergent sums and integrals are computed using Euler–Maclaurin summation with a regularizing function (e.g., exponential regularization) not so anomalous as $|\omega_n|^{-s}$ in the above.[20]

### 4.4.1 More recent theory

Casimir's analysis of idealized metal plates was generalized to arbitrary dielectric and realistic metal plates by Lifshitz and his students.[21][22] Using this approach, complications of the bounding surfaces, such as the modifications to the Casimir force due to finite conductivity, can be calculated numerically using the tabulated complex dielectric functions of the bounding materials. Lifshitz' theory for two metal plates reduces to Casimir's idealized $1/a^4$ force law for large separations $a$ much greater than the skin depth of the metal, and conversely reduces to the $1/a^3$ force law of the London dispersion force (with a coefficient called a Hamaker constant) for small $a$, with a more complicated dependence on $a$ for intermediate separations determined by the dispersion of the materials.[23]

Lifshitz' result was subsequently generalized to arbitrary multilayer planar geometries as well as to anisotropic and magnetic materials, but for several decades the calculation of Casimir forces for non-planar geometries remained limited to a few idealized cases admitting analytical solutions.[24] For example, the force in the experimental sphere–plate geometry was computed with an approximation (due to Derjaguin) that the sphere radius $R$ is much larger than the separation $a$, in which case the nearby surfaces are nearly parallel and the parallel-plate result can be adapted to obtain an approximate $R/a^3$ force (neglecting both skin-depth and higher-order curvature effects).[24][25] However, in the 2000s a number of authors developed and demonstrated a variety of numerical techniques, in many cases adapted from classical computational electromagnetics, that are capable of accurately calculating Casimir forces for arbitrary geometries and materials, from simple finite-size effects of finite plates to more complicated phenomena arising for patterned surfaces or objects of various shapes.[24][26]

## 4.5 Measurement

One of the first experimental tests was conducted by Marcus Sparnaay at Philips in Eindhoven (Netherlands), in 1958, in a delicate and difficult experiment with parallel plates, obtaining results not in contradiction with the Casimir theory,[27][28] but with large experimental errors. Some of the experimental details as well as some background information on how Casimir, Polder and Sparnaay arrived at this point[29] are highlighted in a 2007 interview with Marcus Sparnaay.

The Casimir effect was measured more accurately in 1997 by Steve K. Lamoreaux of Los Alamos National Laboratory,[30] and by Umar Mohideen and Anushree Roy of the University of California, Riverside.[31] In practice, rather than using two parallel plates, which would require phenomenally accurate alignment to ensure they were parallel, the experiments use one plate that is flat and another plate that is a part of a sphere with a large radius.

In 2001, a group (Giacomo Bressi, Gianni Carugno, Roberto Onofrio and Giuseppe Ruoso) at the University of Padua (Italy) finally succeeded in measuring the Casimir force between parallel plates using microresonators.[32]

## 4.6 Regularisation

In order to be able to perform calculations in the general case, it is convenient to introduce a regulator in the summations. This is an artificial device, used to make the sums finite so that they can be more easily manipulated, followed by the taking of a limit so as to remove the regulator.

The heat kernel or exponentially regulated sum is

$$\langle E(t) \rangle = \frac{1}{2} \sum_n \hbar |\omega_n| \exp(-t|\omega_n|)$$

where the limit $t \to 0^+$ is taken in the end. The divergence of the sum is typically manifested as

$$\langle E(t) \rangle = \frac{C}{t^3} + \text{finite}$$

for three-dimensional cavities. The infinite part of the sum is associated with the bulk constant $C$ which *does not* depend on the shape of the cavity. The interesting part of the sum is the finite part, which is shape-dependent. The Gaussian regulator

$$\langle E(t) \rangle = \frac{1}{2} \sum_n \hbar |\omega_n| \exp(-t^2 |\omega_n|^2)$$

is better suited to numerical calculations because of its superior convergence properties, but is more difficult to use in theoretical calculations. Other, suitably smooth, regulators may be used as well. The zeta function regulator

$$\langle E(s) \rangle = \frac{1}{2} \sum_n \hbar |\omega_n| |\omega_n|^{-s}$$

is completely unsuited for numerical calculations, but is quite useful in theoretical calculations. In particular, divergences show up as poles in the complex $s$ plane, with the bulk divergence at $s = 4$. This sum may be analytically continued past this pole, to obtain a finite part at $s = 0$.

Not every cavity configuration necessarily leads to a finite part (the lack of a pole at $s = 0$) or shape-independent infinite parts. In this case, it should be understood that additional physics has to be taken into account. In particular, at extremely large frequencies (above the plasma frequency), metals become transparent to photons (such as X-rays), and dielectrics show a frequency-dependent cutoff as well. This frequency dependence acts as a natural regulator. There are a variety of bulk effects in solid state physics, mathematically very similar to the Casimir effect, where the cutoff frequency comes into explicit play to keep expressions finite. (These are discussed in greater detail in *Landau and Lifshitz*, "Theory of Continuous Media".)

## 4.7   Generalities

The Casimir effect can also be computed using the mathematical mechanisms of functional integrals of quantum field theory, although such calculations are considerably more abstract, and thus difficult to comprehend. In addition, they can be carried out only for the simplest of geometries. However, the formalism of quantum field theory makes it clear that the vacuum expectation value summations are in a certain sense summations over so-called "virtual particles".

More interesting is the understanding that the sums over the energies of standing waves should be formally understood as sums over the eigenvalues of a Hamiltonian. This allows atomic and molecular effects, such as the van der Waals force, to be understood as a variation on the theme of the Casimir effect. Thus one considers the Hamiltonian of a system as a function of the arrangement of objects, such as atoms, in configuration space. The change in the zero-point energy as a function of changes of the configuration can be understood to result in forces acting between the objects.

In the chiral bag model of the nucleon, the Casimir energy plays an important role in showing the mass of the nucleon is independent of the bag radius. In addition, the spectral asymmetry is interpreted as a non-zero vacuum expectation value of the baryon number, cancelling the topological winding number of the pion field surrounding the nucleon.

## 4.8   Dynamical Casimir effect

The dynamical Casimir effect is the production of particles and energy from an accelerated *moving mirror*. This reaction was predicted by certain numerical solutions to quantum mechanics equations made in the 1970s.[33] In May 2011 an announcement was made by researchers at the Chalmers University of Technology, in Gothenburg, Sweden, of the detection of the dynamical Casimir effect. In their experiment, microwave photons were generated out of the vacuum in a superconducting microwave resonator. These researchers used a modified SQUID to change the effective length of the resonator in time, mimicking a mirror moving at the required relativistic velocity. If confirmed this would be the first experimental verification of the dynamical Casimir effect.[34] [35]

### 4.8.1   Analogies

A similar analysis can be used to explain Hawking radiation that causes the slow "evaporation" of black holes (although this is generally visualized as the escape of one particle from a virtual particle-antiparticle pair, the other particle having been captured by the black hole).

Constructed within the framework of quantum field theory in curved spacetime, the dynamical Casimir effect has been used to better understand acceleration radiation such as the Unruh effect.

### 4.8.2   Dynamical Casimir effect and the Big Bang

From a Scalar field model of the Big Bang the Geometric phase is used in a paper by O'Brien "Dynamical Casimir effect and the Big Bang"[36] to relate the Hartle-Hawking State, Guth's and Linde's Inflationary Model and Baryon asymmetry into a single model.

Using the premise of the Feynman–Stueckelberg interpretation that antiparticles travel backwards in time to reflect off the infinite potential boundary of the Scalar field. The Scalar field undergoes a Dynamical Casimir effect during an adiabatic transition, this adiabatic transition is cyclic and satisfies the requirement of a cyclic adiabatic process. This provides a mechanism for matter-antimatter asymmetry and the formation of real on-mass matter, and all takes place as the universe inflates (Guth & Linde mechanism) from Euclidean geometry to a Minkowski spacetime satisfying the Hartle-Hawking State.

## 4.9   Repulsive forces

There are few instances wherein the Casimir effect can give rise to repulsive forces between uncharged objects. Evgeny Lifshitz showed (theoretically) that in certain circumstances (most commonly involving liquids), repulsive forces can arise.[37] This has sparked interest in applications of the Casimir effect toward the development of levitating devices. An experimental demonstration of the Casimir-based repulsion predicted by Lifshitz was recently carried out by Munday et al.[38] Other scientists have also suggested the use of gain media to achieve a similar levitation effect,[39] though this is controversial because these materials seem to violate fundamental causality constraints and the requirement of thermodynamic equilibrium (Kramers-Kronig relations). Casimir and Casimir-Polder repulsion can in fact occur for sufficiently anisotropic electrical bodies; for a review of the issues involved with repulsion see Milton et al.[40]

## 4.10   Applications

It has been suggested that the Casimir forces have application in nanotechnology,[41] in particular silicon integrated circuit technology based micro- and nanoelectromechanical systems, silicon array propulsion for space drives, and so-called Casimir oscillators.[42]

The Casimir effect shows that quantum field theory allows the energy density in certain regions of space to be negative relative to the ordinary vacuum energy, and it has been shown theoretically that quantum field theory allows states where the energy can be *arbitrarily* negative at a given point,[43] Many physicists such as Stephen Hawking,[44] Kip Thorne,[45] and others[46][47][48] therefore argue that such effects might make it possible to stabilize a traversable wormhole. Similar suggestions have been made for the Alcubierre Drive.

On 4 June 2013 it was reported[49] that a conglomerate of scientists from Hong Kong University of Science and Technology, University of Florida, Harvard University, Massachusetts Institute of Technology, and Oak Ridge National Laboratory have for the first time demonstrated a compact integrated silicon chip that can measure the Casimir force.[50]

## 4.11  See also

- Casimir pressure
- Negative energy
- Scharnhorst effect
- Van der Waals force
- Squeezed vacuum

## 4.12  References

[1] Cyriaque Genet, Francesco Intravaia, Astrid Lambrecht and Serge Reynaud (2004) "Electromagnetic vacuum fluctuations, Casimir and Van der Waals forces"

[2] The Force of Empty Space, Physical Review Focus, 3 December 1998

[3] A. Lambrecht, The Casimir effect: a force from nothing, *Physics World*, September 2002.

[4] American Institute of Physics News Note 1996

[5] Jaffe, R. (2005). "Casimir effect and the quantum vacuum". *Physical Review D* **72** (2): 021301. arXiv:hep-th/0503158. Bibcode:2005PhRvD..72b1301J. doi:10.1103/PhysRevD.72.021301.

[6] Photo of ball attracted to a plate by Casimir effect

[7] "The Casimir effect: a force from nothing". physicsworld.com. 1 September 2002. Retrieved 17 July 2009.

[8] Astrid Lambrecht,Serge Reynaud and Cyriaque Genet" Casimir In The Nanoworld"

[9] Griffiths, D. J.; Ho, E. (2001). "Classical Casimir effect for beads on a string". *American Journal of Physics* **69** (11): 1173. Bibcode:2001AmJPh..69.1173G. doi:10.1119/1.1396620.

[10] Cooke, J. H. (1998). "Casimir force on a loaded string". *American Journal of Physics* **66** (7): 569–561. Bibcode:1998AmJPh..66..569C. doi:10.1119/1.18907.

[11] Denardo, B. C.; Puda, J. J.; Larraza, A. S. (2009). "A water wave analog of the Casimir effect". *American Journal of Physics* **77** (12): 1095. Bibcode:2009AmJPh..77.1095D. doi:10.1119/1.3211416.

[12] Larraza, A. S.; Denardo, B. (1998). "An acoustic Casimir effect". *Physics Letters A* **248** (2–4): 151. Bibcode:1998PhLA..248..151L. doi:10.1016/S0375-9601(98)00652-5.

[13] E. L. Losada" Functional Approach to the Fermionic Casimir Effect"

[14] Michael Bordag, Galina Leonidovna Klimchitskaya, Umar Mohideen (2009). "Chapter I; §3: Field quantization and vacuum energy in the presence of boundaries". *Advances in the Casimir effect*. Oxford University Press. pp. 33 *ff*. ISBN 0-19-923874-X.

[15] SE Rugh, H Zinkernagel; Zinkernagel (2002). "The quantum vacuum and the cosmological constant problem". *Studies in History and Philosophy of Science Part B: Studies in History and Philosophy of Modern Physics* **33** (4): 663–705. doi:10.1016/S1355-2198(02)00033-3.

[16] Bianchi, Eugenio; Rovelli, Carlo (2010). "Why all these prejudices against a constant?". arXiv:1002.3966 [astro-ph.CO].

[17] R.L.Jaffe (2005). "The Casimir Effect and the Quantum Vacuum". arXiv:hep-th/0503158v1.

[18] Simpson, W.M.R. (2014). "Ontological aspects of the Casimir effect". *Studies in History and Philosophy of Science Part B* **48**: 84–88. doi:10.1016/j.shpsb.2014.08.001.

[19] For a brief summary, see the introduction in Passante, R.; Spagnolo, S. (2007). "Casimir-Polder interatomic potential between two atoms at finite temperature and in the presence of boundary conditions". *Physical Review A* **76** (4): 042112. arXiv:0708.2240. Bibcode:2007PhRvA..76d2112P. doi:10.1103/PhysRevA.76.042112.

[20] Ruggiero, Zimerman, and Villani, "Application of Analytic Regularization to the Casimir Forces " Revista Brasileira de Física, Vol. 7, Nº 3, 1977; http://www.sbfisica.org.br/bjp/download/v07/v07a43.pdf

[21] Dzyaloshinskii, I E; Lifshitz, E M; Pitaevskii, Lev P (1961). "GENERAL THEORY OF VAN DER WAALS' FORCES". *Soviet Physics Uspekhi* **4** (2): 153. Bibcode:1961SvPhU...4..153D. doi:10.1070/PU1961v004n02ABEH003330.

[22] Dzyaloshinskii, I E; Kats, E I (2004). "Casimir forces in modulated systems". *Journal of Physics: Condensed Matter* **16** (32): 5659. arXiv:cond-mat/0408348. Bibcode:2004JPCM...16.5659D. doi:10.1088/0953-8984/16/32/003.

[23] V. A. Parsegian, *Van der Waals Forces: A Handbook for Biologists, Chemists, Engineers, and Physicists* (Cambridge Univ. Press, 2006).

[24] Rodriguez, A. W.; Capasso, F.; Johnson, Steven G. (2011). "The Casimir effect in microstructured geometries". *Nature Photonics* **5** (4): 211–221. Bibcode:2011NaPho...5..211R. doi:10.1038/nphoton.2011.39. Review article.

[25] B. V. Derjaguin, I. I. Abrikosova, and E. M. Lifshitz, *Quarterly Reviews, Chemical Society*, vol. 10, 295–329 (1956).

[26] Reid, M. T. H.; White, J.; Johnson, S. G. (2011). "Computation of Casimir interactions between arbitrary three-dimensional objects with arbitrary material properties". *Physical Review A* **84** (1): 010503(R). arXiv:1010.5539. Bibcode:2011PhRvA..84a0503R. doi:10.1103/PhysRevA.84.010503.

[27] Sparnaay, M. J. (1957). "Attractive Forces between Flat Plates". *Nature* **180** (4581): 334. Bibcode:1957Natur.180..334S. doi:10.1038/180334b0.

[28] Sparnaay, M (1958). "Measurements of attractive forces between flat plates". *Physica* **24** (6–10): 751. Bibcode:1958Phy....24..751S. doi:10.1016/S0031-8914(58)80090-7.

[29] Movie

[30] Lamoreaux, S. K. (1997). "Demonstration of the Casimir Force in the 0.6 to 6 μm Range". *Physical Review Letters* **78**: 5. Bibcode:1997PhRvL..78....5L. doi:10.1103/PhysRevLett.78.5.

[31] Mohideen, U.; Roy, Anushree (1998). "Precision Measurement of the Casimir Force from 0.1 to 0.9 μm". *Physical Review Letters* **81** (21): 4549. arXiv:physics/9805038. Bibcode:1998PhRvL..81.4549M. doi:10.1103/PhysRevLett.81.4549.

[32] Bressi, G.; Carugno, G.; Onofrio, R.; Ruoso, G. (2002). "Measurement of the Casimir Force between Parallel Metallic Surfaces". *Physical Review Letters* **88** (4): 041804. arXiv:quant-ph/0203002. Bibcode:2002PhRvL..88d1804B. doi:10.1103/PhysRevLett.88.041804. PMID 11801108.

[33] Fulling, S. A.; Davies, P. C. W. (1976). "Radiation from a Moving Mirror in Two Dimensional Space-Time: Conformal Anomaly". *Proceedings of the Royal Society A* **348** (1654): 393. Bibcode:1976RSPSA.348..393F. doi:10.1098/rspa.1976.0045.

[34] "First Observation of the Dynamical Casimir Effect". Technology Review.

[35] Wilson, C. M.; Johansson, G.; Pourkabirian, A.; Simoen, M.; Johansson, J. R.; Duty, T.; Nori, F.; Delsing, P. (2011). "Observation of the Dynamical Casimir Effect in a Superconducting Circuit". *Nature* **479** (7373): 376–379. arXiv:1105.4714. Bibcode:2011Natur.479..376W. doi:10.1038/nature10561. PMID 22094697.

[36] J.C.O'Brien "" (2014)

[37] Dzyaloshinskii, I.E.; Lifshitz, E.M.; Pitaevskii, L.P. (1961). "The general theory of van der Waals forces†". *Advances in Physics* **10** (38): 165. Bibcode:1961AdPhy..10..165D. doi:10.1080/00018736100101281.

[38] Munday, J.N.; Capasso, F.; Parsegian, V.A. (2009). "Measured long-range repulsive Casimir-Lifshitz forces". *Nature* **457** (7226): 170–3. Bibcode:2009Natur.457..170M. doi:10.1038/nature07610. PMID 19129843.

[39] Highfield, Roger (6 August 2007). "Physicists have 'solved' mystery of levitation". *The Daily Telegraph* (London). Retrieved 28 April 2010.

[40] Milton, K. A.; Abalo, E. K.; Parashar, Prachi; Pourtolami, Nima; Brevik, Iver; Ellingsen, Simen A. (2012). "Repulsive Casimir and Casimir-Polder Forces". *J. Phys. A* **45** (37): 4006. arXiv:1202.6415v2. Bibcode:2012JPhA...45K4006M. doi:10.1088/1751-8113/45/37/374006.

[41] Capasso, F.; Munday, J.N.; Iannuzzi, D.; Chan, H.B. (2007). "Casimir forces and quantum electrodynamical torques: physics and nanomechanics". *IEEE Journal of Selected Topics in Quantum Electronics* **13** (2): 400. doi:10.1109/JSTQE.2007.893082.

[42] Serry, F.M.; Walliser, D.; MacLay, G.J. (1995). "The anharmonic Casimir oscillator (ACO)-the Casimir effect in a model microelectromechanical system" (PDF). *Journal of Microelectromechanical Systems* **4** (4): 193. doi:10.1109/84.475546.

[43] Everett, Allen; Roman, Thomas (2012). *Time Travel and Warp Drives.* University of Chicago Press. p. 167. ISBN 0-226-22498-8.

[44] "Space and Time Warps". Hawking.org.uk. Retrieved 2010-11-11.

[45] Morris, Michael; Thorne, Kip; Yurtsever, Ulvi (1988). "Wormholes, Time Machines, and the Weak Energy Condition" (PDF). *Physical Review Letters* **61** (13): 1446–1449. Bibcode:1988PhRvL..61.1446M. doi:10.1103/PhysRevLett.61.1446. PMID 10038800.

[46] Sopova; Ford (2002). "The Energy Density in the Casimir Effect". *Physical Review D* **66** (4): 045026. arXiv:quant-ph/0204125. Bibcode:2002PhRvD..66d5026S. doi:10.1103/PhysRevD.66.045026.

[47] Ford; Roman (1995). "Averaged Energy Conditions and Quantum Inequalities". *Physical Review D* **51** (8): 4277–4286. arXiv:gr-qc/9410043. Bibcode:1995PhRvD..51.4277F. doi:10.1103/PhysRevD.51.4277.

[48] Olum (1998). "Superluminal travel requires negative energies". *Physical Review Letters* **81** (17): 3567–3570. arXiv:gr-qc/9805003. Bibcode:1998PhRvL..81.3567O. doi:10.1103/PhysRevLett.81.3567.

[49] "Chip harnesses mysterious Casimir effect force". 4 June 2013. Retrieved 4 June 2013.

[50] Zao, J.; Marcet, Z.; Rodriguez, A. W.; Reid, M. T. H.; McCauley, A. P.; Kravchenko, I. I.; Lu, T.; Bao, Y.; Johnson, S. G.; Chan, H. B.; et al. (14 May 2013). "Casimir forces on a silicon micromechanical chip". *Nature Communications* **4**: 1845. arXiv:1207.6163. Bibcode:2013NatCo...4E1845Z. doi:10.1038/ncomms2842. PMID 23673630. Retrieved 5 June 2013.

## 4.13   Further reading

### 4.13.1   Introductory readings

- Casimir effect description from University of California, Riverside's version of the Usenet physics FAQ.

- A. Lambrecht, The Casimir effect: a force from nothing, *Physics World*, September 2002.

- Casimir effect on Astronomy Picture of the Day

### 4.13.2   Papers, books and lectures

- H. B. G. Casimir, and D. Polder, "The Influence of Retardation on the London-van der Waals Forces", *Phys. Rev.* **73**, 360–372 (1948).

- Casimir, H. B. G. (1948). "On the attraction between two perfectly conducting plates". *Proc. Kon. Nederland. Akad. Wetensch.* **B51**: 793–795.

- S. K. Lamoreaux, "Demonstration of the Casimir Force in the 0.6 to 6 μm Range", *Phys. Rev. Lett.* **78**, 5–8 (1997)

- M. Bordag, U. Mohideen, V.M. Mostepanenko, "New Developments in the Casimir Effect", *Phys. Rep.* **353**, 1–205 (2001), arXiv. *(200+ page review paper.)*

- Kimball A.Milton: "The Casimir effect", World Scientific, Singapore 2001,ISBN 981-02-4397-9

- Diego Dalvit, et al.: *Casimir Physics.* Springer, Berlin 2011, ISBN 978-3-642-20287-2

- Bressi, G.; Carugno, G.; Onofrio, R.; Ruoso, G. (2002). "Measurement of the Casimir Force between Parallel Metallic Surfaces". *Physical Review Letters* **88** (4): 041804. arXiv:quant-ph/0203002. Bibcode:2002PhRvL..88d1804B. doi:10.1103/PhysRevLett.88.041804. PMID 11801108.

- Kenneth, O.; Klich, I.; Mann, A.; Revzen, M. (2002). "Repulsive Casimir Forces". *Physical Review Letters* **89** (3): 033001. arXiv:quant-ph/0202114. Bibcode:2002PhRvL..89c3001K. doi:10.1103/PhysRevLett.89.033001. PMID 12144387.

- J. D. Barrow, "Much ado about nothing", (2005) Lecture at Gresham College. *(Includes discussion of French naval analogy.)*

- Barrow, John D. (2000). *The book of nothing: vacuums, voids, and the latest ideas about the origins of the universe* (1st American ed.). New York: Pantheon Books. ISBN 0-09-928845-1. (Also includes discussion of French naval analogy.)

- Jonathan P. Dowling, "The Mathematics of the Casimir Effect", *Math. Mag.* **62**, 324–331 (1989).

- Patent № PCT/RU2011/000847 Author Urmatskih.

### 4.13.3 Temperature dependence

- Measurements Recast Usual View of Elusive Force from NIST

- V.V. Nesterenko, G. Lambiase, G. Scarpetta, Calculation of the Casimir energy at zero and finite temperature: some recent results, arXiv:hep-th/0503100 v2 13 May 2005

## 4.14 External links

- Casimir effect article search on arxiv.org

- G. Lang, The Casimir Force web site, 2002

- J. Babb, bibliography on the Casimir Effect web site, 2009

- Paper by J.C.O'Brien (2014). : Dynamical Casimir effect and the Big Bang

# Chapter 5

# Earnshaw's theorem

**Earnshaw's Theorem** states that a collection of point charges cannot be maintained in a stable stationary equilibrium configuration solely by the electrostatic interaction of the charges. This was first proven by British mathematician Samuel Earnshaw in 1842. It is usually referenced to magnetic fields, but was first applied to electrostatic fields.

Earnshaw's theorem applies to classical inverse-square law forces (electric and gravitational) and also to the magnetic forces of permanent magnets, if the magnets are hard (the magnets do not vary in strength with external fields). Earnshaw's theory forbids magnetic levitation in many common situations.

If the materials are not hard, Braunbeck's extension shows that materials with relative magnetic permeability greater than one (paramagnetism) are further destabilising, but materials with a permeability less than one (diamagnetic materials) permit stable configurations.

## 5.1 Explanation

Informally, the case of a point charge in an arbitrary static electric field is a simple consequence of Gauss's law. For a particle to be in a stable equilibrium, small perturbations ("pushes") on the particle in any direction should not break the equilibrium; the particle should "fall back" to its previous position. This means that the force field lines around the particle's equilibrium position should all point inwards, towards that position. If all of the surrounding field lines point towards the equilibrium point, then the divergence of the field at that point must be negative (i.e. that point acts as a sink). However, Gauss's Law says that the divergence of any possible electric force field is zero in free space. In mathematical notation, an electrical force $\mathbf{F}(\mathbf{r})$ deriving from a potential $U(\mathbf{r})$ will always be divergenceless (satisfy Laplace's equation):

$$\nabla \cdot \mathbf{F} = \nabla \cdot (-\nabla U) = -\nabla^2 U = 0.$$

Therefore, there are no local minima or maxima of the field potential in free space, only saddle points. A stable equilibrium of the particle cannot exist and there must be an instability in at least one direction.

To be completely rigorous, strictly speaking, the existence of a stable point does not require that all neighboring force vectors point exactly toward the stable point; the force vectors could spiral in towards the stable point, for example. One method for dealing with this invokes the fact that, in addition to the divergence, the curl of any electric field in free space is also zero (in the absence of any magnetic currents).

It is also possible to prove this theorem directly from the force/energy equations for static magnetic dipoles (below). Intuitively, though, it's plausible that if the theorem holds for a single point charge then it would also hold for two opposite point charges connected together. In particular, it would hold in the limit where the distance between the charges is decreased to zero while maintaining the dipole moment - that is, it would hold for an electric dipole. But if the theorem holds for an electric dipole then it will also hold for a magnetic dipole since the (static) force/energy equations take the same form for both electric and magnetic dipoles.

As a practical consequence, then, this theorem also states that there is no possible static configuration of ferromagnets which can stably levitate an object against gravity, even when the magnetic forces are stronger than the gravitational forces.

Earnshaw's theorem has even been proven for the general case of extended bodies, and this is so even if they are flexible and conducting, provided they are not diamagnetic,[1][2] as diamagnetism constitutes a (small) repulsive force, but no attraction.

There are, however, several exceptions to the rule's assumptions which allow magnetic levitation.

## 5.2 Loopholes

Earnshaw's theorem has no exceptions for non-moving permanent ferromagnets. However, Earnshaw's theorem does not necessarily apply to moving ferromagnets,[3] certain electromagnetic systems, pseudo-levitation and diamagnetic materials. These can thus seem to be exceptions, though in fact they exploit the constraints of the theorem.

Spinning ferromagnets (such as the Levitron) can—while spinning—magnetically levitate using only permanent ferromagnets.[3] Note that since this is spinning, this is not a non-moving ferromagnet.

Switching the polarity of an electromagnet or system of electromagnets can levitate a system by continuous expenditure of energy. Maglev trains are one application.

Pseudo-levitation constrains the movement of the magnets usually using some form of a tether or wall. This works because the theorem shows only that there is some direction in which there will be an instability. Limiting movement in that direction allows for levitation with fewer than the full 3 dimensions available for movement (note that the theorem is proven for 3 dimensions, not 1D or 2D).

Diamagnetic materials are excepted because they exhibit only repulsion against the magnetic field, whereas the theorem requires materials that have both repulsion and attraction. An example of this is the famous levitating frog (see diamagnetism).

## 5.3 Effect on physics

Configurations of classical charged particles orbiting one another are unstable due to losses of energy by electromagnetic radiation. Even without those losses, Earnshaw's theorem means that dynamic systems of charges are unstable over long periods . For quite some time, this led to the puzzling question of why matter stays together as much evidence was found that matter was held together electromagnetically, but static configurations would be unstable, and electrodynamic configurations would be expected to radiate energy and decay.

These questions eventually pointed the way to quantum mechanical explanations of the structure of the atom, and it turns out that the Pauli exclusion principle and the existence of discrete electron orbitals is responsible for making bulk matter rigid.

## 5.4 Proofs for magnetic dipoles

### 5.4.1 Introduction

While a more general proof may be possible, three specific cases are considered here. The first case is a magnetic dipole of constant magnitude that has a fast (fixed) orientation. The second and third cases are magnetic dipoles where the orientation changes to remain aligned either parallel or antiparallel to the field lines of the external magnetic field. In paramagnetic and diamagnetic materials the dipoles are aligned parallel and antiparallel to the field lines, respectively.

### 5.4.2   Background

The proofs considered here are based on the following principles.

The energy U of a magnetic dipole with a magnetic dipole moment $\mathbf{M}$ in an external magnetic field $\mathbf{B}$ is given by

$$U = -\mathbf{M} \cdot \mathbf{B} = -(M_x B_x + M_y B_y + M_z B_z).$$

The dipole will only be stably levitated at points where the energy has a minimum. The energy can only have a minimum at points where the Laplacian of the energy is greater than zero. That is, where

$$\nabla^2 U = \frac{\partial^2 U}{\partial x^2} + \frac{\partial^2 U}{\partial y^2} + \frac{\partial^2 U}{\partial z^2} > 0.$$

Finally, because both the divergence and the curl of a magnetic field are zero (in the absence of current or a changing electric field), the Laplacians of the individual components of a magnetic field are zero. That is,

$$\nabla^2 B_x = \nabla^2 B_y = \nabla^2 B_z = 0.$$

This is proven at the very end of this article as it is central to understanding the overall proof.

### 5.4.3   Summary of proofs

For a magnetic dipole of fixed orientation (and constant magnitude) the energy will be given by

$$U = -\mathbf{M} \cdot \mathbf{B} = -(M_x B_x + M_y B_y + M_z B_z),$$

where $Mx$, $My$ and $Mz$ are constant. In this case the Laplacian of the energy is always zero,

$$\nabla^2 U = 0,$$

so the dipole can have neither an energy minimum nor an energy maximum. That is, there is no point in free space where the dipole is either stable in all directions or unstable in all directions.

Magnetic dipoles aligned parallel or antiparallel to an external field with the magnitude of the dipole proportional to the external field will correspond to paramagnetic and diamagnetic materials respectively. In these cases the energy will be given by

$$U = -\mathbf{M} \cdot \mathbf{B} = -k\mathbf{B} \cdot \mathbf{B} = -k\left(B_x^2 + B_y^2 + B_z^2\right),$$

where $k$ is a constant greater than zero for paramagnetic materials and less than zero for diamagnetic materials.

In this case, it will be shown that

$$\nabla^2 \left(B_x^2 + B_y^2 + B_z^2\right) \geq 0,$$

which, combined with the constant $k$, shows that paramagnetic materials can have energy maxima but not energy minima and diamagnetic materials can have energy minima but not energy maxima. That is, paramagnetic materials can be unstable in all directions but not stable in all directions and diamagnetic materials can be stable in all directions but not unstable in all directions. Of course, both materials can have saddle points.

Finally, the magnetic dipole of a ferromagnetic material (a permanent magnet) that is aligned parallel or antiparallel to a magnetic field will be given by

$$\mathbf{M} = k\frac{\mathbf{B}}{|\mathbf{B}|},$$

so the energy will be given by

$$U = -\mathbf{M} \cdot \mathbf{B} = -k\frac{\mathbf{B} \cdot \mathbf{B}}{|\mathbf{B}|} = -k\frac{|\mathbf{B}|^2}{|\mathbf{B}|} = -k\left(B_x^2 + B_y^2 + B_z^2\right)^{\frac{1}{2}};$$

but this is just the square root of the energy for the paramagnetic and diamagnetic case discussed above and, since the square root function is monotonically increasing, any minimum or maximum in the paramagnetic and diamagnetic case will be a minimum or maximum here as well. There are, however, no known configurations of permanent magnets that stably levitate so there may be other reasons not discussed here why it is not possible to maintain permanent magnets in orientations antiparallel to magnetic fields (at least not without rotation—see Levitron).

### 5.4.4 Detailed proofs

Earnshaw's theorem was originally formulated for electrostatics (point charges) to show that there is no stable configuration of a collection of point charges. The proofs presented here for individual dipoles should be generalizable to collections of magnetic dipoles because they are formulated in terms of energy, which is additive. A rigorous treatment of this topic is, however, currently beyond the scope of this article.

### 5.4.5 Fixed-orientation magnetic dipole

It will be proven that at all points in free space

$$\nabla \cdot (\nabla U) = \nabla^2 U = \frac{\partial^2 U}{\partial x^2} + \frac{\partial^2 U}{\partial y^2} + \frac{\partial^2 U}{\partial z^2} = 0.$$

The energy $U$ of the magnetic dipole $\mathbf{M}$ in the external magnetic field $\mathbf{B}$ is given by

$$U = -\mathbf{M} \cdot \mathbf{B} = -M_x B_x - M_y B_y - M_z B_z.$$

The Laplacian will be

$$\nabla^2 U = -\frac{\partial^2 (M_x B_x + M_y B_y + M_z B_z)}{\partial x^2} - \frac{\partial^2 (M_x B_x + M_y B_y + M_z B_z)}{\partial y^2} - \frac{\partial^2 (M_x B_x + M_y B_y + M_z B_z)}{\partial z^2}.$$

Expanding and rearranging the terms (and noting that the dipole $\mathbf{M}$ is constant) we have

$$\nabla^2 U = -M_x\left(\frac{\partial^2 B_x}{\partial x^2} + \frac{\partial^2 B_x}{\partial y^2} + \frac{\partial^2 B_x}{\partial z^2}\right) - M_y\left(\frac{\partial^2 B_y}{\partial x^2} + \frac{\partial^2 B_y}{\partial y^2} + \frac{\partial^2 B_y}{\partial z^2}\right) - M_z\left(\frac{\partial^2 B_z}{\partial x^2} + \frac{\partial^2 B_z}{\partial y^2} + \frac{\partial^2 B_z}{\partial z^2}\right)$$
$$= -M_x \nabla^2 B_x - M_y \nabla^2 B_y - M_z \nabla^2 B_z$$

but the Laplacians of the individual components of a magnetic field are zero in free space (not counting electromagnetic radiation) so

$$\nabla^2 U = -M_x 0 - M_y 0 - M_z 0 = 0,$$

which completes the proof.

### 5.4.6   Magnetic dipole aligned with external field lines

The case of a paramagnetic or diamagnetic dipole is considered first. The energy is given by

$$U = -k|\mathbf{B}|^2 = -k\left(B_x^2 + B_y^2 + B_z^2\right).$$

Expanding and rearranging terms,

$$\nabla^2 |\mathbf{B}|^2 = \nabla^2 \left(B_x^2 + B_y^2 + B_z^2\right)$$
$$= 2\left(|\nabla B_x|^2 + |\nabla B_y|^2 + |\nabla B_z|^2 + B_x \nabla^2 B_x + B_y \nabla^2 B_y + B_z \nabla^2 B_z\right)$$

but since the Laplacian of each individual component of the magnetic field is zero,

$$\nabla^2 |\mathbf{B}|^2 = 2\left(|\nabla B_x|^2 + |\nabla B_y|^2 + |\nabla B_z|^2\right);$$

and since the square of a magnitude is always positive,

$$\nabla^2 |\mathbf{B}|^2 \geq 0.$$

As discussed above, this means that the Laplacian of the energy of a paramagnetic material can never be positive (no stable levitation) and the Laplacian of the energy of a diamagnetic material can never be negative (no instability in all directions).

Further, because the energy for a dipole of fixed magnitude aligned with the external field will be the square root of the energy above, the same analysis applies.

### 5.4.7   Laplacian of individual components of a magnetic field

It is proven here that the Laplacian of each individual component of a magnetic field is zero. This shows the need to invoke the properties of magnetic fields that the divergence of a magnetic field is always zero and the curl of a magnetic field is zero in free space. (That is, in the absence of current or a changing electric field.) See Maxwell's equations for a more detailed discussion of these properties of magnetic fields.

Consider the Laplacian of the x component of the magnetic field

$$\nabla^2 B_x = \frac{\partial^2 B_x}{\partial x^2} + \frac{\partial^2 B_x}{\partial y^2} + \frac{\partial^2 B_x}{\partial z^2}$$
$$= \frac{\partial}{\partial x}\frac{\partial B_x}{\partial x} + \frac{\partial}{\partial y}\frac{\partial B_x}{\partial y} + \frac{\partial}{\partial z}\frac{\partial B_x}{\partial z}$$

Because the curl of $\mathbf{B}$ is zero,

$$\frac{\partial B_x}{\partial y} = \frac{\partial B_y}{\partial x},$$

and

$$\frac{\partial B_x}{\partial z} = \frac{\partial B_z}{\partial x},$$

so we have

$$\nabla^2 B_x = \frac{\partial}{\partial x}\frac{\partial B_x}{\partial x} + \frac{\partial}{\partial y}\frac{\partial B_y}{\partial x} + \frac{\partial}{\partial z}\frac{\partial B_z}{\partial x}.$$

But since $Bx$ is continuous, the order of differentiation doesn't matter giving

$$\nabla^2 B_x = \frac{\partial}{\partial x}\left(\frac{\partial B_x}{\partial x} + \frac{\partial B_y}{\partial y} + \frac{\partial B_z}{\partial z}\right) = \frac{\partial}{\partial x}(\nabla \cdot \mathbf{B}).$$

The divergence of **B** is zero,

$$\nabla \cdot \mathbf{B} = 0,$$

so

$$\nabla^2 B_x = \frac{\partial}{\partial x}(\nabla \cdot \mathbf{B}) = 0.$$

The Laplacian of the $y$ component of the magnetic field $By$ field and the Laplacian of the $z$ component of the magnetic field $Bz$ can be calculated analogously. Alternatively, one can use the identity

$$\nabla^2 \mathbf{B} = \nabla(\nabla \cdot \mathbf{B}) - \nabla \times (\nabla \times \mathbf{B}),$$

where both terms in the parentheses vanish.

## 5.5 Notes

[1] Gibbs, Philip & Geim, Andre. "Is Magnetic Levitation Possible?". High Field Magnet Laboratory. Retrieved 2010-01-04.

[2] Earnshaw, S., On the nature of the molecular forces which regulate the constitution of the luminferous ether., Trans. Camb. Phil. Soc., 7, pp 97-112 (1842)

[3] American Journal of Physics, April 1997 Spin stabilized magnetic levitation Martin D. Simon, UCLA Department of Physics

**References**

- Earnshaw, Samuel (1842). "On the Nature of the Molecular Forces which Regulate the Constitution of the Luminiferous Ether". *Trans. Camb. Phil. Soc.* **7**: 97–112.

- Scott, W. T. (1959). "Who Was Earnshaw?". *American Journal of Physics* **27**: 418. Bibcode:1959AmJPh..27..418S. doi:10.1119/1.1934886.

## 5.6 See also

- Magnetic levitation
- Electrostatic levitation

## 5.7 External links

- "Is magnetic levitation possible?", a discussion of Earnshaw's theorem and its consequences for levitation, along with several ways to levitate with electromagnetic fields

- Biography and other information about Samuel Earnshaw

# Chapter 6

# Electrogravitics

Not to be confused with Gravitoelectromagnetism, in which gravity behaves similarly to magnetism.

**Electrogravitics** is claimed to be an unconventional type of effect or anti-gravity propulsion created by an electric field's effect on a mass. The name was coined in the 1920s by the discoverer of the effect, Thomas Townsend Brown, who spent most of his life trying to develop it and sell it as a propulsion system. Through Brown's promotion of the idea it was researched for a short while by aerospace companies in the 1950s. Electrogravitics is popular with conspiracy theorists with claims that it is powering flying saucers and the B-2 Stealth Bomber.

Since apparatus based on Browns' ideas have often yielded varying and highly controversial results when tested within controlled vacuum conditions, the effect observed has often been attributed to the ion drift or ion wind effect instead of anti-gravity.[1]

Brown also named this phenomenon the "Biefeld–Brown effect" after his claimed mentor, Denison University professor Paul Alfred Biefeld.

## 6.1   Origins

Electrogravitics had its origins in experiments started in 1921 by Thomas Townsend Brown (USA) (who coined the name) while he was still in high school. He discovered an unusual effect while experimenting with a Coolidge tube, a type of X-ray vacuum tube where, if he placed on a balance scale with the tube's positive electrode facing up, the tubes mass seemed to decrease, when facing down the tube's mass seemed to increase.[2] Brown showed this effect to his college professors and even newspaper reporters and told them he was convinced that he had managed to influence gravity electronically. Brown developed this into large high voltage capacitors that would produce a tiny propulsive force causing the capacitor to jump in one direction when the power was turned on. In 1929 Brown published "How I Control Gravity," in *Science and Invention* where he claimed the capacitors were producing a mysterious force that interacted with the pull of gravity. He envisions a future where, if his device could be scaled up, "Multi-impulse gravitators weighing hundreds of tons may propel the ocean liners of the future" or even "fantastic 'space cars'" to Mars.[3] Somewhere along the way Brown came up with the name Biefeld–Brown effect, named after his former teacher, professor of astronomy Paul Alfred Biefeld at Denison University in Ohio. Brown claimed Biefeld as his mentor and co-experimenter.[4][5] After World War II Brown sought to develop the effect as a means of propulsion for aircraft and spacecraft, demonstrating a working apparatus to an audience of scientists and military officials in 1952. Research in the phenomenon was popular in the mid-1950s, at one point the Glenn L. Martin Company placed advertisements looking for scientists who were "interested in gravity", but rapidly declined in popularity thereafter.

Instead of being an anti-gravity force, this effect has been found to be caused by ionized particles exerting a force between two asymmetrical electrodes that produces a type of ion drift or ionic wind that transfers its momentum to surrounding neutral particles, an electrokinetic phenomena or more widely referred to as *electrohydrodynamics* (EHD).[6]

## 6.2 Claims

Electrogravitics has become popular with UFO, anti-gravity, and government conspiracy theorists[3] where it is seen as an example of something much more exotic than electrokinetics, i.e. that electrogravitics is a true anti-gravity technology that can "create a force that depends upon an object's mass, even as gravity does".[7][8] There are claims that all major aerospace companies in the 1950s including Martin, Convair, Lear, Sperry, Raytheon were working on it, that the technology became highly classified in the early 1960s, that it is used to power the B-2 bomber,[3] and that it can be used to generate "free energy".[9] Charles Berlitz devoted an entire chapter of his book on The Philadelphia Experiment (*The Philadelphia Experiment: Project Invisibility*) to a retelling of Brown's early work with the effect, implying the electrogravitics effect was being used by UFOs. The researcher and author Paul LaViolette has produced many self-published books on electrogravitics, making many claims over the years including his view that the technology could have helped to avoid another Space Shuttle Columbia disaster.

Nikola Tesla has also been associated with electrogravitics.[9]

### 6.2.1 Criticism

Many claims as to the validity of electrogravitics as an anti-gravity force revolve around research and videos on the internet purported to show lifter-style capacitor devices working in a vacuum, therefor not receiving propulsion from ion drift or ion wind being generated in air.[3][10] Followups on the claims (R. L. Talley in a 1990 US Air Force study, NASA scientist Jonathan Campbell in a 2003 experiment,[11] and Martin Tajmar in a 2004 paper[12]) have found that no thrust could be observed in a vacuum, consistent with the phenomenon of ion wind. Campbell pointed out to a Wired magazine reporter that creating a true vacuum similar to space for the test requires tens of thousands of dollars in equipment.

Byron Preiss in his 1985 book on the current science and future of the Solar System titled *The Planets* commented that electrogravitics development seemed to be "much ado about nothing, started by a bunch of engineers who didn't know enough physics". Preiss stated that electrogravitics, like exobiology, is "a science without a single specimen for study".[13]

## 6.3 See also

- United States gravity control propulsion initiative
- List of topics characterized as pseudoscience

## 6.4 References

[1] Stein, W.B. 2000: Electrokinetic Propulsion: The Ion Wind Argument. Purdue University, Energy Conversion Lab (Hangar #3, Purdue Airport, West Lafayette, IN 47906)

[2] The Canonical Hamiltonian The Intersection Of Chip Design and Physics by Hamilton Carter, Thomas Townsend Brown: Part IV of the Holiday Serial

[3] Thompson, Clive (August 2003). "The Antigravity Underground". *Wired Magazine*.

[4] Paul Schatzkin, Defying Gravity: The Paraellel Universe of T. Townsend Brown, 2005-2006-2007-2008 - Tanglewood Books, Chapter 13: Notes from the Rabbit Hole #3: "He Made Things Up" (online excerpts)

[5] alienscientist.com, Biefeld-Brown Effect Controversy, Tajmar ESA Experiments

[6] NASA CR-2004-213312 Asymmetrical Capacitors for propulsion

[7] Thomas F. Valone, Progress in Electrogravitics and Electrokinetics for Aviation and Space Travel - Integrity Research Institute, Washington DC

[8] activistpost.com, Sunday, April 1, 2012 Electrogravitics – A Simplified Description, Amaterasu Solar

[9] Chapter Six UFOs and Electrogravity Propulsion, Did Tesla Discover the Secrets of Antigravity?

[10] Thomas Valone, Electrogravitics II: Validating Reports on a New Propulsion Methodology, Integrity Research Institute, page 52-58

[11] Thompson, Clive (August 2003). "The Antigravity Underground". *Wired Magazine*.

[12] Tajmar, M. (2004). "Biefeld-Brown Effect: Misinterpretation of Corona Wind Phenomena". *AIAA Journal* **42** (2): 315. Bibcode:2004AIAAJ..42..315T. doi:10.2514/1.9095.

[13] Byron Preiss (1985). *The Planets*. Bantam Books. p. 27. ISBN 0-553-05109-1.

## 6.5   Further reading

- Thomas Valone, *Electrogravitics Systems: Reports on a New Propulsion Methodology*. Integrity Research Institute; 2nd ed edition (November 1995). 102 pages. ISBN 0-9641070-0-7 ISBN 978-0964107007

- Thomas Valone, *Electrogravitics II: Validating Reports on a New Propulsion Methodology*. Integrity Research Institue; 2Rev Ed edition (July 1, 2005). 160 pages. ISBN 0-9641070-9-0 ISBN 978-0964107090

- Jen-shih Chang, *Handbook of Electrostatic Processes*. CRC Press, 1995. ISBN 0-8247-9254-8

- Nick Cook, *The Hunt for Zero Point: Inside the Classified World of Antigravity Technology*. Broadway; 1 edition (August 13, 2002). 304 pages ISBN 0-7679-0627-6 ISBN 978-0767906272

- Paul A. LaViolette, "Secrets of Antigravity Propulsion: Tesla, UFOs, and Classified Aerospace Technology". Bear & Company, Rochester VT (2008), Paperback: 512 pages, ISBN 978-1-59143-078-0

## 6.6   External links

- Electrogravitics at American Antigravity A page of YouTube talks and demonstrations by supporters.

- UFO How-To Volume II: Electrogravitics Excerpts from the book.

# Chapter 7

# Electrostatic levitation

**Electrostatic levitation** is the process of using an electric field to levitate a charged object and counteract the effects of gravity. It was used, for instance, in Robert Millikan's oil drop experiment and is used to suspend the gyroscopes in Gravity Probe B during launch.

Due to Earnshaw's theorem no static arrangement of classical electrostatic fields can be used to stably levitate a point charge. There is an equilibrium point where the two fields cancel, but it is an unstable equilibrium. By using feedback techniques it is possible to adjust the charges to achieve a quasi static levitation.

## 7.1 Earnshaw's theorem

Main article: Earnshaw's theorem

The idea of particle instability in an electrostatic field originated with Samuel Earnshaw in 1839 [1] and was formalized by James Clerk Maxwell [2] in 1874 who gave it the title "Earnshaw's theorem" and proved it with the Laplace equation. Earnshaw's theorem explains why a system of electrons is not stable and was invoked by Niels Bohr in his atom model of 1913[3] when criticizing J. J. Thomson's atom.

Earnshaw's theorem holds that a charged particle suspended in an electrostatic field is unstable, because the forces of attraction and repulsion vary at an equal rate that is proportional to the inverse square law and remain in balance wherever a particle moves. Since the forces remain in balance, there is no inequality to provide a restoring force; and the particle remains unstable and can freely move without restriction.

## 7.2 Levitation

The first electrostatic levitator was invented by Dr. Won-Kyu Rhim at NASA's JPL lab in 1993.[4] A charged sample of 2 mm in diameter can be levitated in a vacuum chamber between two electrodes positioned vertically with an electrostatic field in between. The field is controlled through a feedback system to keep the levitated sample at a predetermined position. Several copies of this system have been made in JAXA and NASA, and the original system has been transferred to California Institute of Technology with an upgraded setup of tetrahedra four beam laser heating system.

On the Moon the photoelectric effect and electrons in the solar wind charge fine layers of moon dust on the surface forming an atmosphere of dust that floats in "fountains" over the surface of the moon.[5][6]

## 7.3 See also

- Magnetic levitation

- Optical levitation

- Acoustic levitation

- Aerodynamic levitation

- Biefeld-Brown effect

- EHD thruster

- Ionocraft (*Lifter*)

## 7.4 References

[1] Samuel Earnshaw "On the Nature of the Molecular Forces which regulate the Constitution of the Luminiferous Ether," *Transactions of the Cambridge Philosophical Society,* Cambridge University Press, Vol. 7, pp. 97-122 (1842).

[2] James Clerk Maxwell, *A Treatise on Electricity and Magnetism,* Macmillan and Co., "Earnshaw's theorem" p. 139 (1873)

[3] Niels Bohr, "On the Constitution of Atoms and Molecules," *The London, Edinburgh, and Dublin Philosophical Magazine and Journal of Science,* Vol. 7 July (1913).

[4] W. K. Rhim, S. K. Chung, D. Barber, K. F. Man, G. Gutt, A. Rulison, and R. E. Spjut, Review of Scientific Instruments 64, 2961 (1993).

[5] Bell, Trudy E., "Moon fountains", FirstScience.com, 2001-01-06.

[6] Dust gets a charge in a vacuum

## 7.5 External links

- JLN Labs: Levitators

- Electrostatic levitator — Marshall Space Flight Center

- Electrostatic levitation raises dust particles off the surface of the moon

- Hybrid electric/acoustic levitation

- Electrostatic levitation and transportation of glass or silicon plates

- Electrostatic levitation of various materials including silicon, cobalt palladium, aluminium and other compounds

*Sample of a titanium-zirconium-nickel alloy inside the Electrostatic Levitator vacuum chamber at NASA's Marshall Space Flight Center.*

# Chapter 8

# Flux pumping

**Flux pumping** is a method for magnetising superconductors to fields in excess of 15 teslas. The method can be applied to any type II superconductor and exploits a fundamental property of superconductors. That is their ability to support and maintain currents on the length scale of the superconductor. Conventional magnetic materials are magnetised on a molecular scale which means that superconductors can maintain a flux density orders of magnitude bigger than conventional materials. Flux pumping is especially significant when one bears in mind that all other methods of magnetising superconductors require application of a magnetic flux density at least as high as the final required field. This is not true of flux pumping.

An electric current flowing in a loop of superconducting wire can persist indefinitely with no power source. In a normal conductor, an electric current may be visualized as a fluid of electrons moving across a heavy ionic lattice. The electrons are constantly colliding with the ions in the lattice, and during each collision some of the energy carried by the current is absorbed by the lattice and converted into heat, which is essentially the vibrational kinetic energy of the lattice ions. As a result, the energy carried by the current is constantly being dissipated. This is the phenomenon of electrical resistance.

The situation is different in a superconductor. In a conventional superconductor, the electronic fluid cannot be resolved into individual electrons. Instead, it consists of bound *pairs* of electrons known as Cooper pairs. This pairing is caused by an attractive force between electrons from the exchange of phonons. Due to quantum mechanics, the energy spectrum of this Cooper pair fluid possesses an *energy gap*, meaning there is a minimum amount of energy $\Delta E$ that must be supplied in order to excite the fluid. Therefore, if $\Delta E$ is larger than the thermal energy of the lattice, given by $kT$, where $k$ is Boltzmann's constant and $T$ is the temperature, the fluid will not be scattered by the lattice. The Cooper pair fluid is thus a superfluid, meaning it can flow without energy dissipation.

In a class of superconductors known as type II superconductors, including all known high-temperature superconductors, an extremely small amount of resistivity appears at temperatures not too far below the nominal superconducting transition when an electric current is applied in conjunction with a strong magnetic field, which may be caused by the electric current. This is due to the motion of vortices in the electronic superfluid, which dissipates some of the energy carried by the current. If the current is sufficiently small, the vortices are stationary, and the resistivity vanishes. The resistance due to this effect is tiny compared with that of non-superconducting materials, but must be taken into account in sensitive experiments.

## 8.1 Introduction

In the method described here a magnetic field is swept across the superconductor in a magnetic wave. This field induces current according to Faraday's law of induction. As long as the direction of motion of the magnetic wave is constant then the current induced will always be in the same sense and successive waves will induce more and more current.

Traditionally the magnetic wave would be generated either by physically moving a magnet or by an arrangement of coils switched in sequence, such as occurs on the stator of a three-phase motor. Flux Pumping is a solid state method where a material which changes magnetic state at a suitable magnetic ordering temperature is heated at its edge and the resultant thermal wave produces a magnetic wave which then magnetizes the superconductor. A superconducting flux pump should not be confused with a classical flux pump as described in Van Klundert et al.'s[1] review.

The method described here has two unique features:

- At no point is the superconductor driven normal; the procedure simply makes modifications to the critical state.

- The critical state is not modified by a moving magnet or an array of solenoids, but by a thermal pulse which modifies the magnetization, thus sweeping vortices into the material.

The system, as described, is actually a novel kind of heat engine in which thermal energy is being converted into magnetic energy.

## 8.2   Background

### 8.2.1   Meissner effect

Main article: Meissner effect
When a superconductor is placed in a weak external magnetic field **H**, the field penetrates the superconductor only a small

*Persistent electric current flows on the surface of the superconductor, acting to exclude the magnetic field of the magnet. This current effectively forms an electromagnet that repels the magnet.*

distance $\lambda$, called the London penetration depth, decaying exponentially to zero within the interior of the material. This is called the Meissner effect, and is a defining characteristic of superconductivity. For most superconductors, the London penetration depth is on the order of 100 nm.

The Meissner effect is sometimes confused with the kind of diamagnetism one would expect in a perfect electrical conductor: according to Lenz's law, when a *changing* magnetic field is applied to a conductor, it will induce an electric current in the conductor that creates an opposing magnetic field. In a perfect conductor, an arbitrarily large current can be induced, and the resulting magnetic field exactly cancels the applied field.

The Meissner effect is distinct from this because a superconductor expels *all* magnetic fields, not just those that are changing. Suppose we have a material in its normal state, containing a constant internal magnetic field. When the material is cooled below the critical temperature, we would observe the abrupt expulsion of the internal magnetic field, which we would not expect based on Lenz's law.

The Meissner effect was explained by the brothers Fritz and Heinz London, who showed that the electromagnetic free energy in a superconductor is minimized provided

$$\nabla^2 \mathbf{H} = \lambda^{-2} \mathbf{H}$$

where $\mathbf{H}$ is the magnetic field and $\lambda$ is the London penetration depth.

This equation, which is known as the London equation, predicts that the magnetic field in a superconductor decays exponentially from whatever value it possesses at the surface.

In 1962, the first commercial superconducting wire, a niobium-titanium alloy, was developed by researchers at Westinghouse, allowing the construction of the first practical superconducting magnets. In the same year, Josephson made the important theoretical prediction that a supercurrent can flow between two pieces of superconductor separated by a thin layer of insulator.[2] This phenomenon, now called the Josephson effect, is exploited by superconducting devices such as SQUIDs. It is used in the most accurate available measurements of the magnetic flux quantum $\Phi_0 = \frac{h}{2e}$ , and thus (coupled with the quantum Hall resistivity) for Planck's constant $h$. Josephson was awarded the Nobel Prize for this work in 1973.

### 8.2.2  E-J Power Law

The most popular model used to describe superconductivity is the Bean or Critical State model and variations such as the Kim-Anderson model. However the Bean model assumes zero resistivity and that current is always induced at the critical current. A more useful model for engineering applications is the so-called E_J power law in which the field and the current are linked by the following equations:

$$\rho = \frac{dE}{dJ}$$

$$\mathbf{E} = \mathbf{E_0} * \left(\frac{J}{J_c}\right)^n$$

$$\rho(J) = \frac{E_0 * n * (\frac{J}{J_c})^{n-1}}{J_c}$$

In these equations if n = 1 then the conductor has linear resistivity such as is found in copper. The higher the n-value the closer we get to the critical state model. Also the higher the n-value then the "better" the superconductor as the lower the resistivity at a certain current. The E_J power law can be used to describe the phenomenon of flux-creep in which a superconductor gradually loses its magnetisation over time. This process is logarithmic and thus gets slower and slower and ultimately leads to very stable fields.

## 8.3  Theory

The potential of superconducting coils and bulk melt-processed YBCO single domains to maintain significant magnetic fields at cryogenic temperatures makes them particularly attractive for a variety of engineering applications including superconducting magnets, magnetic bearings and motors. It has already been shown that large fields can be obtained in

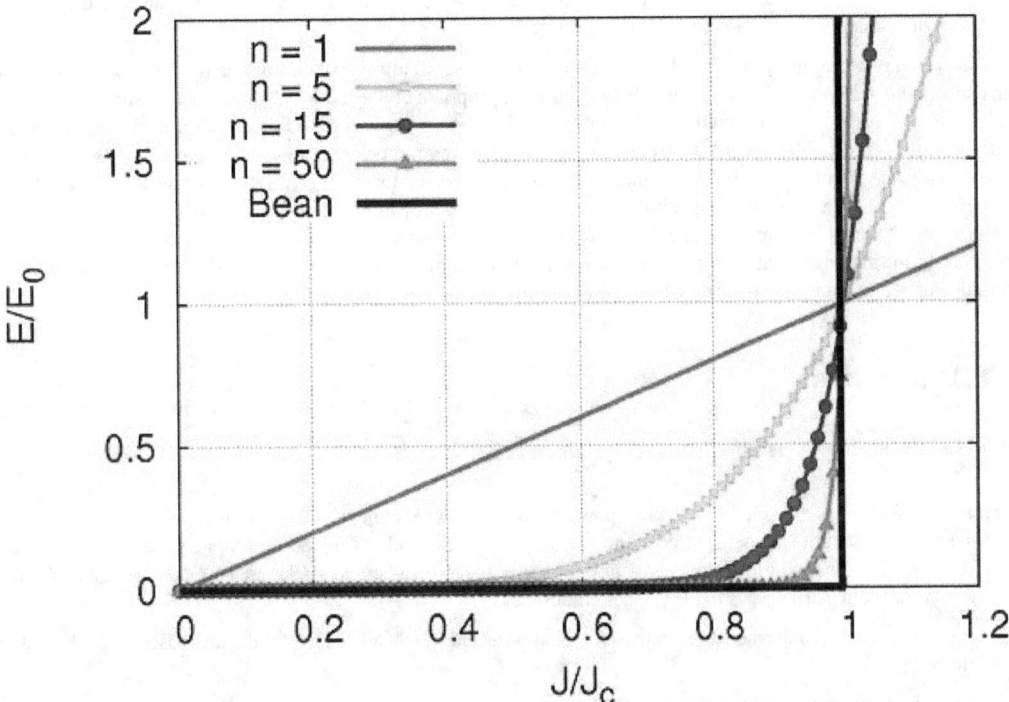

single domain bulk samples at 77 K. A range of possible applications exist in the design of high power density electric motors.

Before such devices can be created a major problem needs to be overcome. Even though all of these devices use a superconductor in the role of a permanent magnet and even though the superconductor can trap potentially huge magnetic fields (greater than 10 T) the problem is the induction of the magnetic fields, this applies both to bulk and to coils operating in persistent mode. There are four possible known methods:

1. Cooling in field;

2. Zero field cooling, followed by slowly applied field;

3. Pulse magnetization;

4. Flux pumping;

Any of these methods could be used to magnetise the superconductor and this may be done either in situ or ex situ. Ideally the superconductors are magnetised in situ.

There are several reasons for this: first, if the superconductors should become demagnetised through (i) flux creep, (ii) repeatedly applied perpendicular fields or (iii) by loss of cooling then they may be re-magnetized without the need to disassemble the machine. Secondly, there are difficulties with handling very strongly magnetized material at cryogenic temperatures when assembling the machine. Thirdly, ex situ methods would require the machine to be assembled both cold and pre-magnetized and would offer significant design difficulties. Until room temperature superconductors can be prepared, the most efficient design of machine will therefore be one in which an in situ magnetizing fixture is included!

The first three methods all require a solenoid which can be switched on and off. In the first method an applied magnetic field is required equal to the required magnetic field, whilst the second and third approaches require fields at least two times greater. The final method, however, offers significant advantages since it achieves the final required field by repeated applications of a small field and can utilise a permanent magnet.

If we wish to pulse a field using, say, a 10 T magnet to magnetize a 30 mm × 10 mm sample then we can work out how big the solenoid needs to be. If it were possible to wind an appropriate coil using YBCO tape then, assuming an $I^c$ of 70 A and a thickness of 100 μm, we would have 100 turns and 7 000 A turns. This would produce a B field of approximately $7\ 000/(20 \times 10^{-3}) \times 4\pi \times 10^{-7} = 0.4$ T. To produce 10 T would require pulsing to 1 400 A! An alternative calculation would be to assume a $J_c$ of say $5 \times 10^8\,\mathrm{Am}^{-1}$ and a coil 1 cm$^2$ in cross section. The field would then be $5 \times 10^8 \times 10^{-2} \times (2 \times 4\pi \times 10^{-7}) = 10$ T. Clearly if the magnetisation fixture is not to occupy more room than the puck itself then a very high activation current would be required and either constraint makes in situ magnetization a very difficult proposition. What is required for in situ magnetisation is a magnetisation method in which a relatively small field of the order of milliteslas repeatedly applied is used to magnetize the superconductor.

## 8.4 Applications

Main article: Technological applications of superconductivity

Superconducting magnets are some of the most powerful electromagnets known. They are used in MRI and NMR machines, mass spectrometers, Magnetohydrodynamic Power Generation and beam-steering magnets used in particle accelerators. They can also be used for magnetic separation, where weakly magnetic particles are extracted from a background of less or non-magnetic particles, as in the pigment industries.

Other early markets are arising where the relative efficiency, size and weight advantages of devices based on HTS outweigh the additional costs involved.

Promising future applications include high-performance transformers, power storage devices, electric power transmission, electric motors (e.g. for vehicle propulsion, as in vactrains or maglev trains), magnetic levitation devices, and fault current limiters.

## 8.5 References

[1] L.J.M. van de Klundert; et al. (1981). "On fully conducting rectifiers and fluxpumps. A review. Part 2: Commutation modes, characteristics and switches". *Cryogencis*: 267–277.

[2] B.D. Josephson (1962). "Possible new effects in superconductive tunnelling". *Phys. Lett.* **1** (7): 251–253. Bibcode:1962PhL.....1..251J. doi:10.1016/0031-9163(62)91369-0.

## 8.6 Sources

- Coombs, Timothy (2008). "Superconductors the next generation of permanent magnets" (PDF).

- Qiuliang Wang et al., "Study of Full-wave Superconducting Rectifier-type Flux-pumps", IEEE Transactions on Magnetics, vol. 32, No. 4, pp. 2699–2702, Jul. 1996.

- Coombs, Timothy (2007). "A novel heat engine for magnetising superconductors" (PDF).

- Coombs, Timothy; Hong, Z; Zhu, X (2007). "A thermally actuated superconducting flux pump" (PDF). *Physica C: Superconductivity* **468** (3): 153. Bibcode:2008PhyC..468..153C. doi:10.1016/j.physc.2007.11.003.

- L.J.M. van de Klundert et al., "On fully conducting rectifiers and fluxpumps. A review. Part 2: Commutation modes, characteristics and switches", Cryogencis, pp. 267–277, May 1981.

- L.J.M. van de Klundert et al., "Fully superconducting rectifiers and fluxpumps Part 1: Realized methods for pumping flux", Cryogenics, pp. 195–206, Apr. 1981.

- Kleinert, Hagen, *Gauge Fields in Condensed Matter*, Vol. I, " SUPERFLOW AND VORTEX LINES"; Disorder Fields, Phase Transitions, pp. 1–742, World Scientific (Singapore, 1989); Paperback ISBN 9971-5-0210-0 *(also readable online: Vol. I)*

- Larkin, Anatoly; Varlamov, Andrei, *Theory of Fluctuations in Superconductors*, Oxford University Press, Oxford, United Kingdom, 2005 (ISBN 0-19-852815-9)

- A.G. Lebed (Ed.) (2008). *The Physics of Organic Superconductors and Conductors* (1st ed.). Springer Series in Materials Science , Vol. 110. ISBN 9783540766728.

- Tinkham, Michael (2004). *Introduction to Superconductivity* (2nd ed.). Dover Books on Physics. ISBN 0-486-43503-2.

- Tipler, Paul; Llewellyn, Ralph (2002). *Modern Physics* (4th ed.). W. H. Freeman. ISBN 0-7167-4345-0.

## 8.7   External links

- Recent publications
- Magnetohydrodynamic Power Generation

# Chapter 9

# Viktor Grebennikov

**Viktor Stepanovich Grebennikov** (Виктор Степанович Гребенников) (1927–2001) was a self-proclaimed Russian scientist, naturalist, entomologist and paranormal researcher best known for his claim to have invented a levitation platform which operated by attaching dead insect body parts to the underside. Grebennikov wrote detailed accounts of his experiences flying over the Russian countryside using his levitation device. These flying experiences as well as his reported observations of other paranormal phenomena, usually involving insect nests or parts, appear in his self-published book *My World* (*Moi Mir*. Novosibirsk, Russia: Sovetskaya Sibir, 1997).[1]

Although once popular with readers who dreamed of human unpowered flight, Grebennikov's flight and other paranormal claims were promptly rejected by skeptics and scientists outside of the paranormal community as his reports were devoid of conclusive proof or public demonstration. He claimed that his camera shutter was jammed during the flights due to a time-warping force-field generated by the secret "geometric" power of chitin.[1]

He was granted a Russian patent in 1993 on a device containing beehive cells (dry honeycomb) that is claimed to enhance the effectiveness of therapeutic drugs in a patient.[2]

The paranormal author Brian Snellgrove cites some of Grebennikov's *My World* claims in his books.[3][4]

## 9.1 References

[1] Grebennikov, Viktor Stepanovich. "Chapter 5: The Natural Phenomena of AntiGravitation and Invisibility in Insects due to the Grebennikov Cavity Structure Effect (CSE)". *My World*. Translation posted on KeelyNet.com.

[2] RU patent 2061509, Grebennikov, Viktor S., "Device For Applying Energy", published 1996, assigned to Chastnoe semejnoe predprijatie Nekrasova-Luchkina "ALNI". Copy with abstract in English at RexResearch.com.

[3] Snellgrove, Brian (1996). *The Unseen Self: Kirlian Photography Explained*. Random House UK. ISBN 9780852072776.

[4] Snellgrove, Brian (2011). "Chapter 3: Perceiving the Aura". *The Magic In Your Hands: How to See Auras and Use Them for Diagnosis and Healing*. Ebury. pp. 31–33. ISBN 9781446447314.

## 9.2 External links

- Гребенникова, Книга Виктора. Мой мир[*My World*] (in Russian).

- Cherednichenko, Juri N. "Victor S. Grebennikov: Insect Chitin Anti-Gravity & Cavity Structural Effect (CSE)". *RexResearch.com*. Jean, NV: Robert A. Nelson.

# Chapter 10

# Ionocraft

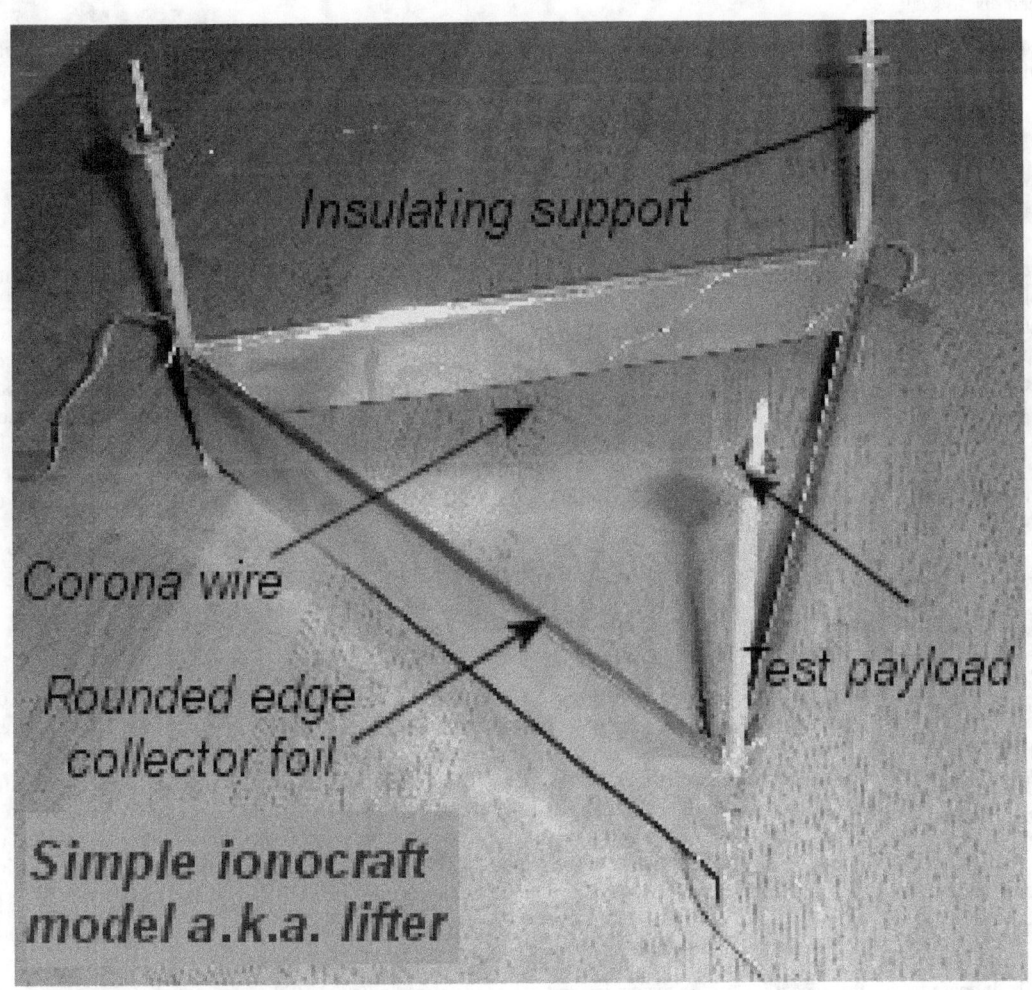

*A simple ionocraft, generating 15 gF (0.147 newtons) of thrust at 54 kV*

An **ionocraft** or **ion-propelled aircraft** (commonly known as a **lifter** or **hexalifter**) is a device that uses an electrical electrohydrodynamic (EHD) phenomenon to produce thrust in the air without requiring any combustion or moving parts.

The term "ionocraft" dates back to the 1960s, an era in which EHD experiments were at their peak. In its basic form, it simply consists of two parallel conductive electrodes, one in the form of a fine wire and another which may be formed out of wire grid, tubes or foil skirts with a smooth round surface. When such an arrangement is powered by high voltage (in the range of a few kilovolts), it produces thrust. The ionocraft forms part of the EHD thruster family, but is a special case in which the ionisation and accelerating stages are combined into a single stage.

The device is a popular science fair project for students. It is also popular among anti-gravity or so-called "electrogravitics" proponents, due to the research of Thomas Townsend Brown, who built these devices in the 1920s and incorrectly believed that he had found a way to modify gravity using electric fields.

The term "lifter" is an accurate description because it is not an anti-gravity device; rather, it produces lift using the same basic principle as a rocket, i.e. from the equal but opposite force upward generated by the driving force downward, specifically by driving the ionized air downward in the case of the ionocraft. Much like a rocket or a jet engine (it can actually be much more thrust efficient than a jet engine[1]), the force that an ionocraft generates is consistently oriented along its own axis, regardless of the surrounding gravitational field. Claims of the device also working in a vacuum have been disproved.[2]

Ionocraft require many safety precautions due to the high voltage required for their operation; nevertheless, a large sub-culture has grown up around this simple EHD thrusting device and its physics are now known to a much better extent.

## 10.1 Description

An ionocraft is a propulsion device based on ionic air propulsion that works without moving parts, uses only electrical energy, and is able to lift its own weight, not including its own power supply. The principle of ionic wind propulsion with corona-generated charged particles has been known from the earliest days of the discovery of electricity with references dating back to 1709 in a book titled *Physico-Mechanical Experiments on Various Subjects* by Francis Hauksbee. Its use for propulsion was given serious thought by Major Alexander Prokofieff de Seversky who contributed much to its basic physics and construction variations in 1960. In fact, it was Major de Seversky himself who in 1964 coined the term Ionocraft in his (U.S. Patent 3,130,945). There are also designs by the American experimenter Thomas Townsend Brown, such as his 1960 patents for "Elektrokinetic Apparatus". Brown spent most of his life trying to develop what he thought was an anti-gravity effect, which he named the Biefeld–Brown effect. Since Brown's devices produce thrust along their axis regardless of the direction of gravity and do not work in a vacuum, the effect he identified has been attributed to electrohydrodynamics instead of anti-gravity.[3][4]

## 10.2 Construction

A simple ionocraft derivative, also known as a lifter, can be easily constructed by anyone with a minimal amount of technical knowledge. The model in its simplest form has the shape of an equilateral triangle with sides generally between 10 and 30 cm. They consist of three parts, the *corona wire* (or emitting wire), the *air gap* (or dielectric fluid), and the *foil skirt* (collector). The electrical polarities of the emitting and collecting electrodes can be reversed. All of this is usually supported by a lightweight balsawood or other electrically isolating frame so that the corona wire is supported at a fixed distance above the foil skirt, generally at 1 mm per kilo-volt. The corona wire and foil should be as close as possible to achieve a saturated corona current condition which results in the highest production of thrust. However the corona wire should not be too close to the foil skirt as it will tend to arc in a spectacular show of tiny lightning bolts which has a twofold effect:

1. It degrades the thrust as it is shorting the device and there is current flow through the arc instead of the ions that do the lifting

2. It can destroy the power supply or burn the balsa structure of the lifter.

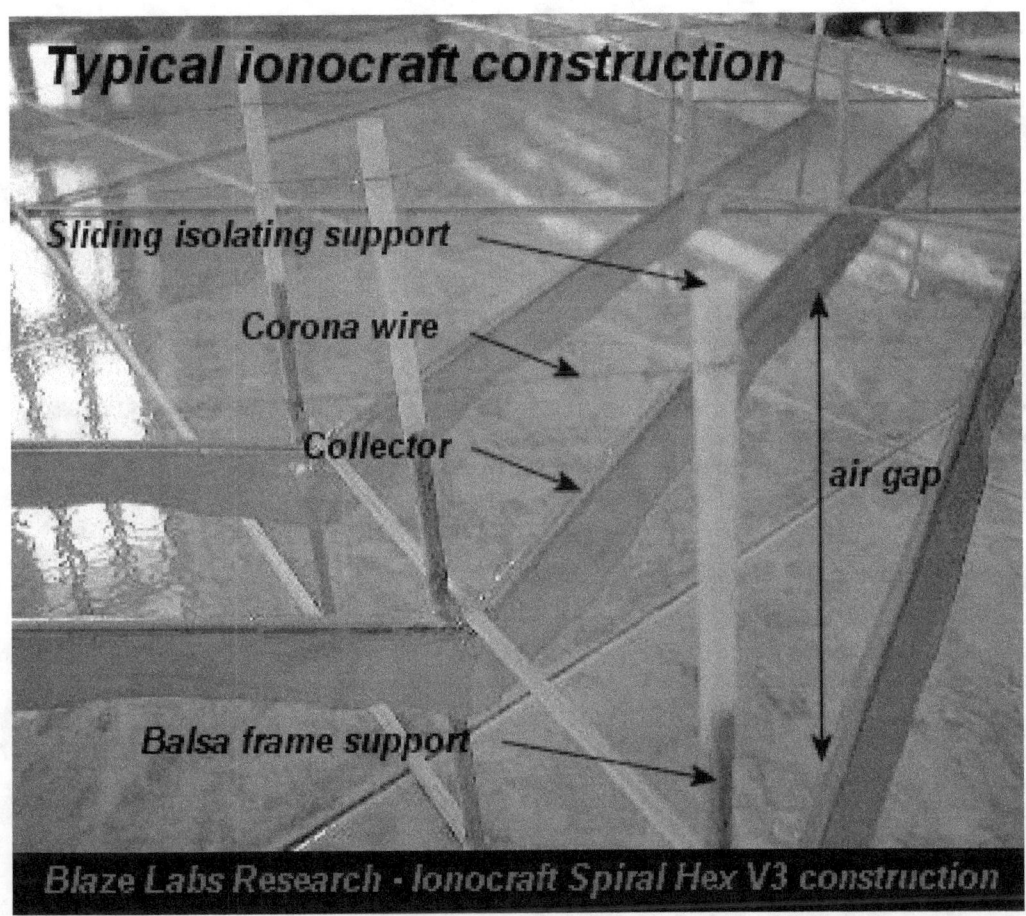

*Typical ionocraft construction*

### 10.2.1   Components

**Corona wire**

The corona wire is usually, but not necessarily, connected to the positive terminal of the high voltage power supply. In general, it is made from a small gauge bare conductive wire. While copper wire can be used, it does not work quite as well as stainless steel. Similarly, thinner wire such as 50 gauge tends to work well compared to more common, larger sizes such as 30 gauge, as the stronger electric field around the smaller diameter wire results in better ionisation and a larger corona current.

The corona wire is so called because of its tendency to emit a purple corona-like glow while in use. This is simply a side effect of ionization. Excessive corona is to be avoided, as too much means the electrodes are dangerously close and may arc at any moment, not to mention the associated health hazards due to excess inhalation of ozone and NOx produced by the corona.

**Air gap**

The air gap is simply that, a gap of free flowing air between the two electrodes that make up the structure of an ionocraft.

The air gap is a vital necessity to the functioning of this device as it is the dielectric used during operation. Best results have been observed with an air gap of 1 mm to every kV.

**Collector**

The collector may take various shapes, as long as it results in a smooth equipotential surface underneath the corona wire. Variations of this include a wire mesh, parallel conductive tubes, or a foil skirt with a smooth round edge. The foil skirt collector is the most popular for small models, and is usually, but not necessarily, connected to the negative side of the power supply. It is usually conveniently made from cheap, lightweight aluminum foil.

The foil skirt is named simply because it is shaped much like a skirt, and is made from aluminum foil. It is by far the most fragile part, and must not be crumpled to work properly. Any sharp edges on the skirt will degrade the performance of the thruster, as this will generate ions of opposite polarity to those within the thrust mechanism.

Reversing the polarities of the corona wire with that of the foil does not alter the direction of motion. Thrust will be produced regardless of whether the ions are positive or negative. For positive corona polarity, nitrogen ions are the main charge carriers, whilst for negative polarity, oxygen ions will be the main carriers and ozone production will be higher. The slight difference in their ion mobility, results in slightly higher thrust for the positive corona polarity case.

## 10.3  Mechanism

The generated thrust can be explained in terms of electrokinetics or, in modern terms, electrohydrodynamics propulsion and can be derived through a modified use of the Child-Langmuir equation.[5]

A generalized one-dimensional treatment gives the equation:

$$F = \frac{Id}{k}$$

where

- $F$ is the resulting force, measured in dimension M L $T^{-2}$

- $I$ is the current flow of electric current, measured in dimension I.

- $d$ is the air gap distance, measured in dimension L.

- $k$ is the ion mobility coefficient of air, measured in dimension $M^{-1}$ $T^2$ I (Nominal value $2 \cdot 10^{-4}$ $m^2$ $V^{-1}$ $s^{-1}$).

In its basic form, the ionocraft is able to produce forces great enough to lift about a gram of payload per watt,[6] so its use is restricted to a tethered model. Ionocraft capable of payloads in the order of a few grams usually need to be powered by power sources and high voltage converters weighing a few kilograms, so although its simplistic design makes it an excellent way to experiment with this technology, it is unlikely that a fully autonomous ionocraft will be made with the present construction methods. Further study in electrohydrodynamics, however, show that different classes and construction methods of EHD thrusters and hybrid technology (mixture with lighter-than-air techniques), can achieve much higher payload or thrust-to-power ratios than those achieved with the simple lifter design. Practical limits can be worked out using well defined theory and calculations.[7] Thus, a fully autonomous EHD thruster is theoretically possible.

When the ionocraft is turned on, the corona wire becomes charged with high voltage, usually between 20 and 50 kV. The user must be extremely careful not to touch the device at this point, as it can give a nasty shock. At extremely high current, well over the amount usually used for a small model, contact could be fatal. When the corona wire is at approximately 30 kV, it causes the air molecules nearby to become ionised by stripping the electrons away from them. As this happens, the ions are strongly repelled away from the anode but are also strongly attracted towards the collector, causing the majority of the ions to begin accelerating in the direction of the collector. These ions travel at a constant average velocity termed the drift velocity. Such velocity depends on the mean free path between collisions, the external electric field, and on the mass of ions and neutral air molecules.

The fact that the current is carried by a corona discharge (and not a tightly-confined arc) means that the moving particles are diffusely spread out into an expanding ion cloud, and collide frequently with neutral air molecules. It is these collisions that create a net movement. The momentum of the ion cloud is partially imparted onto the neutral air molecules that it collides with, which, being neutral, do not eventually migrate back to the second electrode. Instead they continue to travel in the same direction, creating a neutral wind. As these neutral molecules are ejected from the ionocraft, there are, in agreement with Newton's Third Law of Motion, equal and opposite forces, so the ionocraft moves in the opposite direction with an equal force. There are hundreds of thousands of molecules per second ejected from the device, so the force exerted is comparable to a gentle breeze. Still, this is enough to make a light balsa model lift its own weight. The resulting thrust also depends on other external factors including air pressure and temperature, gas composition, voltage, humidity, and air gap distance.

The air gap is very important for the function of this device. Between the electrodes there is a mass of air, consisting of neutral air molecules, which gets in the way of the moving ions. This air mass is impacted repeatedly by excited particles moving at high drift velocity. This creates resistance, which must be overcome. The barrage of ions will eventually either push the whole mass of air out of the way, or break through to the collector where electrons will be reattached, making it neutral again. The end result of the neutral air caught in the process is to effectively cause an exchange in momentum and thus generate thrust. The heavier and denser the gas, the higher the resulting thrust.

Recent research suggests electrohydrodynamic propulsion is more energy efficient (thrust per unit power) than other means of propulsion, generating up to 100N of thrust per kilowatt of power, compared to 2 N/kW for jet engines.[8] This is mainly due to the much lower air speed of an ionocraft vs a jet engine, as power requirement per unit mass of payload drops with air velocity. However this also means the ionocraft needs a much wider surface area to lift the same payload.

## 10.4   See also

- Alexander Prokofieff de Seversky

- Electrohydrodynamics

- Hall effect thruster

- Ion thruster

- Magnetoplasmadynamic thruster

- Plasma actuator

- Wingless Electromagnetic Air Vehicle

## 10.5   References

[1]  Massachusetts Institute of Technology (2013, April 3). Ionic thrusters generate efficient propulsion in air. ScienceDaily Quote: "...In their experiments, they found that ionic wind produces 110 newtons of thrust per kilowatt, compared with a jet engine's 2 newtons per kilowatt..."

[2]  "Ion Propulsion" (PDF).

[3]  Thompson, Clive (August 2003). "The Antigravity Underground". *Wired Magazine*.

[4]  Tajmar, M. (2004). "Biefeld-Brown Effect: Misinterpretation of Corona Wind Phenomena". *AIAA Journal* **42** (2): 315. Bibcode:2004AIAAJ..42..315T. doi:10.2514/1.9095.

[5]  "Electrokinetic devices in air" (PDF). Retrieved 2013-04-25.

[6]  Lifter efficiency relation to ion velocity "J L Naudin's Lifter-3 pulsed HV 1.13g/Watt"

[7] Full analysis & design solutions for EHD Thrusters at saturated corona current conditions

[8] Barrett, Stephen R.H.; Masuyama, Kento (5 March 2013). "On the performance of electrohydrodynamic propulsion". *Proceedings of the Royal Society*. Retrieved 3 April 2013.

## 10.6   Sources

- Talley, R .L., "*Twenty First Century Propulsion Concept*". PLTR-91-3009, Final Report for the period Feb 89 to July 90, on Contract FO4611-89-C-0023, Phillips Laboratory, Air Force Systems Command, Edwards AFB, CA 93523-5000, 1991.

- Tajmar, M., "*Experimental Investigation of 5-D Divergent Currents as a Gravity-Electromagnetism Coupling Concept*". Proceedings of the Space Technology and Applications International Forum (STAIF-2000), El-Genk editor, AIP Conference Proceedings 504, American Institute of Physics, New York, pp. 998–1003, 2000.

- Tajmar, M., "*The Biefeld-Brown Effect: Misinterpretation of Corona Wind Phenomena*". AIAA Journal, Vol 42, pp 315–318 2004.

- DR Buehler, *Exploratory Research on the Phenomenon of the Movement of High Voltage Capacitors*. Journal of Space Mixing, 2004

- FX Canning, C Melcher, E Winet, *Asymmetrical Capacitors for Propulsion*. 2004.

- GV Stephenson *The Biefeld Brown Effect and the Global Electric Circuit*. AIP Conference Proceedings, 2005.

## 10.7   External links

- Electrostatic Antigravity on NASA's "Common Errors in propulsion" page

- NASA: Asymmetrical Capacitors for Propulsion

- http://www.nasa.gov/centers/glenn/technology/Ion_Propulsion1.html

# Chapter 11

# Optical levitation

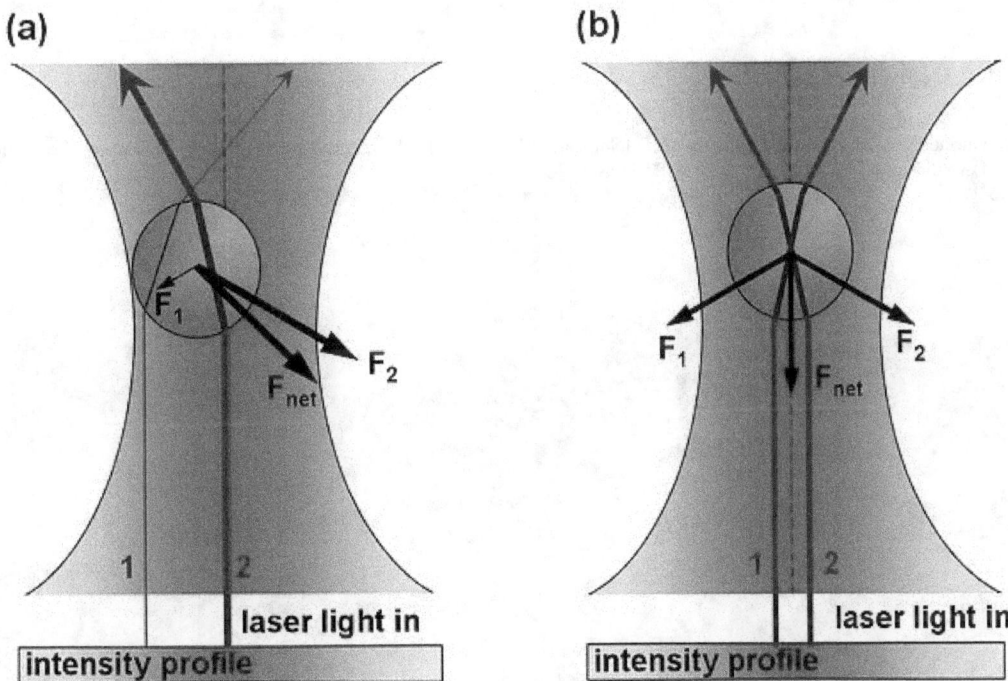

*A force diagram showing how vertical and lateral stabilization occurs in a vertically oriented optical trap.*

**Optical levitation** is a method developed by Arthur Ashkin whereby a material is levitated against the downward force of gravity by an upward force stemming from photon momentum transfer. Typically photon radiation pressure of a vertical upwardly directed and focused laser beam of enough intensity counters the downward force of gravity to allow for a stable *optical trap* capable of holding small particles in suspension.

Micrometer sized (from several to 50 micrometer in diameter) transparent dielectric spheres such as fused silica spheres, oil or water droplets, are used in this type of experiment. The laser radiation can be fixed in wavelength such as that of an argon ion laser or that of a tunable dye laser. Laser power required is of the order of 1 Watt focused to a spot size of several tens of micrometers. Phenomena related to morphology-dependent resonances in a spherical optical cavity have been studied by several research groups.

For a shiny object, such as a metallic micro-sphere, stable optical levitation has not been achieved. Optical levitation of a macroscopic object is also theoretically possible.[1]

## 11.1  See also

- Optical tweezers
- Electrostatic levitation
- Magnetic levitation
- Acoustic levitation
- Aerodynamic levitation
- Laser propulsion
- List of laser articles

## 11.2  References

[1] Guccione, G.; M. Hosseini; S. Adlong; M. T. Johnsson; J. Hope; B. C. Buchler; P. K. Lam (July 2013). "Scattering-Free Optical Levitation of a Cavity Mirror". arXiv:1307.1175.

# Chapter 12

# Squeezed vacuum

**Squeezed vacuum** is an approach to generate negative energy density (NED) [1]

## 12.1  See also

- Casimir effect

## 12.2  References

[1] http://arxiv.org/ftp/arxiv/papers/1005/1005.5682.pdf Emitting solitonized laser beams to boost the negative energy density of squeezed regions of the vacuum

# Chapter 13

# Superdiamagnetism

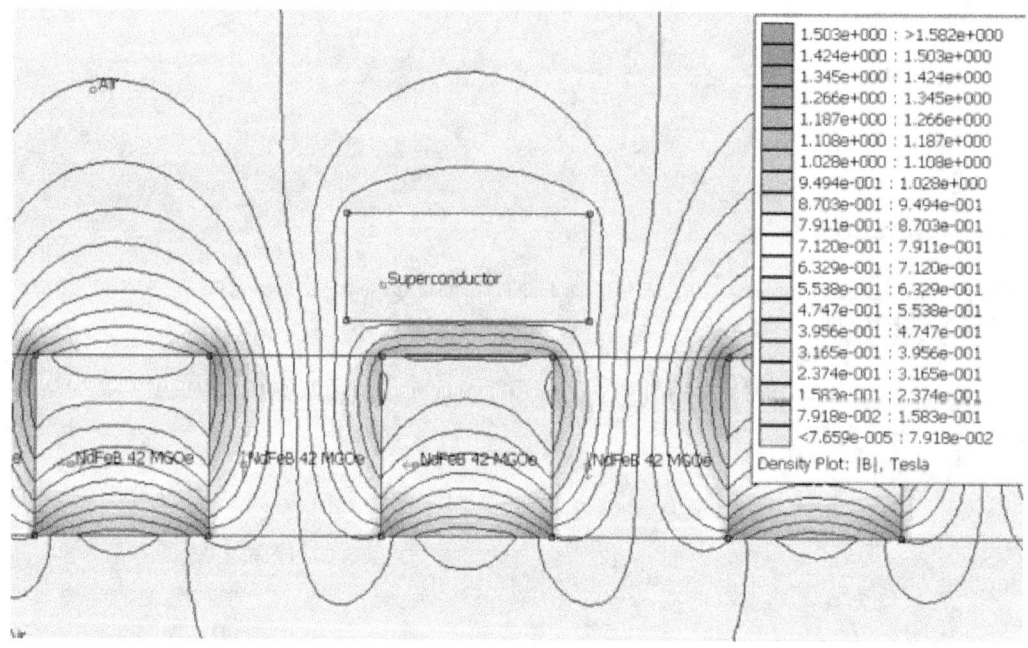

*A superconductor acts as an essentially perfect diamagnetic material when placed in a magnetic field and it excludes the field, and so the flux lines completely avoid the region*

**Superdiamagnetism** (or **perfect diamagnetism**) is a phenomenon occurring in certain materials at low temperatures, characterised by the complete absence of magnetic permeability (i.e. a magnetic susceptibility $\chi_v = -1$) and the exclusion of the interior magnetic field.

Superdiamagnetism established that the superconductivity of a material was a stage of phase transition. Superconducting magnetic levitation is due to superdiamagnetism, which repels a permanent magnet which approaches the superconductor, and flux pinning, which prevents the magnet floating away.

Superdiamagnetism is a feature of superconductivity. It was identified in 1933, by Walther Meissner and Robert Ochsenfeld, but it is considered distinct from the Meissner effect which occurs when the superconductivity first forms, and involves the exclusion of magnetic fields that already penetrate the object.

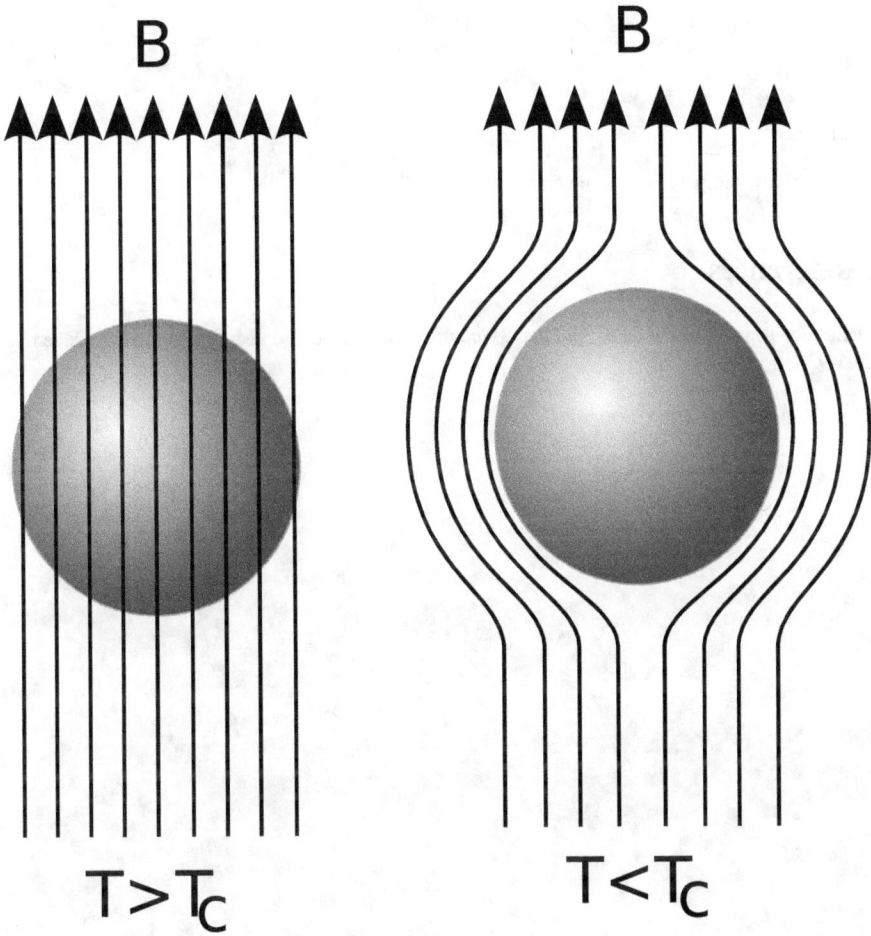

*Diagram of the Meissner effect. Magnetic field lines, are excluded from a superconductor when it is below its critical temperature.*

## 13.1   Theory

Fritz London and Heinz London developed the theory that the exclusion of magnetic flux is brought about by electrical screening currents that flow at the surface of the superconducting material and which generate a magnetic field that exactly cancels the externally applied field inside the superconductor. These screening currents are generated whenever a superconducting material is brought inside a magnetic field. This can be understood by the fact that a superconductor has zero electrical resistance, so that "eddy currents", induced by the motion of the material inside a magnetic field, will not decay. Fritz, at the Royal Society in 1935, stated that the thermodynamic state would be described by a single wave function.

"Screening currents" also appear in a situation wherein an initially normal, conducting metal is placed inside a magnetic field. As soon as the metal is cooled below the appropriate transition temperature, it becomes superconducting. This expulsion of magnetic field upon the cooling of the metal cannot be explained any longer by merely assuming zero resistance and is called *the Meissner effect*. It shows that the superconducting state does not depend on the history of preparation, only upon the present values of temperature, pressure and magnetic field, and therefore is a true thermodynamic state.

## 13.2   See also

- Superfluidity
- Timeline of low-temperature technology

## 13.3   References

- Shachtman, Tom, *Absolute Zero: And the Conquest of Cold*. Houghton Mifflin Company, December 1999. ISBN 0-395-93888-0

## 13.4 Text and image sources, contributors, and licenses

### 13.4.1 Text

- **Levitation** *Source:* https://en.wikipedia.org/wiki/Levitation?oldid=676398199 *Contributors:* Arpingstone, Rainer Wasserfuhr~enwiki, Steinsky, Omegatron, David.Monniaux, Finlay McWalter, Shantavira, Jondel, Michael Snow, Wolfkeeper, Art Carlson, Everyking, No Guru, Mackerm, Utcursch, Andycjp, LucasVB, Beland, Superborsuk, Karol Langner, Ukexpat, Eep$^2$, Discospinster, Rich Farmbrough, RJHall, Project2501a, NickW, Peter Greenwell, Sriram sh, Thialfi, JavOs, Alansohn, Wdfarmer, Sobolewski, Velella, Omphaloscope, Nuno Tavares, Alvis, Eleassar777, Rchamberlain, Rjwilmsi, Quiddity, The wub, Fish and karate, FlaBot, Diza, Antimatter15, Chobot, YurikBot, RobotE, RussBot, Netscott, Chensiyuan, Stassats, Brandon, Zythe, Bhumiya, Vicarious, Kungfuadam, Chic happens, Yvwv, SmackBot, Levitate, KelleyCook, The great kawa, Gilliam, Ohnoitsjamie, Skizzik, Hraefen, Bluebot, Headwes, MalafayaBot, SchfiftyThree, Maxwhale66, Johndemers, Kittybrewster, Das cool, DMacks, Ollj, Carnby, TStone, Theologist, Joffeloff, Cointel, Iridescent, Rfwebster, Audiosmurf, Tawkerbot2, Hyphen5, MrFish, A876, Crossmr, Gogo Dodo, Optimist on the run, Protious, Thijs!bot, MatheMezzaMorphis163, Headbomb, Aristox, MichaelMaggs, Farosdaughter, Yellowdesk, Serpent's Choice, XyBot, Dan D. Ric, HTGuru, .anacondabot, VoABot II, Nyq, Otherjohn, Hikarunogo, Algorythmic, Cgingold, Lawrie.skinner, Baseracer, Ghableska, Spyzo, MartinBot, Xantolus, Creol, Jack franklin 1, J.delanoy, Berserkerz Crit, Wimox, Cometstyles, Alyssa hoffel, Badn3wz~enwiki, Andy Marchbanks, Venomkill112, JohnBlackburne, Craobh sidhe, Epson291, Jake123456789123, Autobogg, Arydberg, LeaveSleaves, Madhero88, Q Science, Red58bill, Closenplay, SieBot, Grundle2600, Oda Mari, Theun k, Lightmouse, Ajbajo01, Rat at WikiFur, ClueBot, Blackangel25, Maxijohn, Polyamorph, Jusdafax, XLinkBot, DazSap, Addbot, DB au, Kont Dracula, Sqib, Gail, Termar, Yobot, Dallasloose, AnomieBOT, Are you ready for IPv6?, Capricorn42, Frankie0607, GorgeCustersSabre, GliderMaven, Paine Ellsworth, Fluteflute123456789, Tom.Reding, DARTH SIDIOUS 2, RjwilmsiBot, RepliCarter, WikiGuy000, Bollyjeff, Teapeat, Rememberway, ClueBot NG, Incompetence, Axhynic, Chester Markel, Ranjbar, Sameeanahmedkhan, Bibcode Bot, Lucuas, Atomician, Oct13, Sazrat17, Mariusz.stepien, StopTheCrackpots, Lugia2453, Petarbosnicpetrus and Anonymous: 174

- **Acoustic levitation** *Source:* https://en.wikipedia.org/wiki/Acoustic_levitation?oldid=659782416 *Contributors:* Michael Hardy, Cimon Avaro, Tom harrison, Anville, LucasVB, Stepp-Wulf, Kghose, DrDeke, BRW, Fadereu, RJFJR, GregorB, CharlesC, Erebus555, Gussisaurio, Chobot, RussBot, Netscott, Gaius Cornelius, SmackBot, Heptite, PointyOintment, TastyPoutine, Asymptote, Headbomb, Kborer, Dvandersluis, Byrgenwulf, Tstrobaugh, Robinresearch, Email4mobile, User A1, Brianonn, Barkeep, Arumugaramprakash, Polyamorph, XLinkBot, Interferometrist, Addbot, Legobot, AnomieBOT, RCraig09, Foxdove, Drdeak, DASHBot, H3llBot, ClueBot NG, Monkbot, Volker Siegel, Danielitos, Anonymous 99922 and Anonymous: 19

- **Aerodynamic levitation** *Source:* https://en.wikipedia.org/wiki/Aerodynamic_levitation?oldid=697794368 *Contributors:* TheParanoidOne, Netscott, Ratarsed, Chairman S., Chris the speller, Bluebot, PointyOintment, Gobonobo, Lawrie.skinner, Ideal gas equation, Polyamorph, Addbot, Yobot, Citation bot, Citation bot 1, Tom.Reding, Braincricket, Bibcode Bot, Fotdjesss and Anonymous: 8

- **Casimir effect** *Source:* https://en.wikipedia.org/wiki/Casimir_effect?oldid=697291247 *Contributors:* Peter Winnberg, Bryan Derksen, XJaM, Heron, Isis~enwiki, JohnOwens, Ixfd64, Loooix~enwiki, Stevenj, Dynabee, Cimon Avaro, Rob Hooft, Gamma~enwiki, Timwi, David Latapie, Evgeni Sergeev, The Anomebot, Wik, Silvonen, Phys, Omegatron, Pakaran, BenRG, Jeffq, Meelar, Intangir, Modeha, Carnildo, Matt Gies, Giftlite, ShaunMacPherson, Varlaam, Soren.harward, Jason Quinn, Adam McMaster, HorsePunchKid, Superborsuk, Karol Langner, DragonflySixtyseven, RetiredUser2, Lumidek, Peter bertok, Rich Farmbrough, Habbit, Pjacobi, Ylai, Bender235, Helldjinn~enwiki, Chewie, JustinWick, Edwinstearns, Orange Julius, Jmah, I9Q79oL78KiL0QTFHgyc, Lysdexia, Eruantalon, Jumbuck, Keenan Pepper, H2g2bob, Dan East, Kazvorpal, AustinZ, Joriki, Richard Arthur Norton (1958- ), Linas, FeanorStar7, Jwanders, Thunderrabbit, Btyner, Christopher Thomas, Scottanon, KyuuA4, Rjwilmsi, Boccobrock, Eck~enwiki, Mishuletz, Ewlyahoocom, Carrionluggage, Diza, Srleffler, Jinma, Wavelength, Angus Lepper, Hairy Dude, Midgley, RussBot, Ytrottier, Gaius Cornelius, Theshadow27, Anomo, Ravedave, Larsobrien, Kortoso, Bota47, Rcinda1, Pace212, Dna-webmaster, Tetracube, WAS 4.250, Enormousdude, 2over0, Geoffrey.landis, Ilmari Karonen, KasugaHuang, SmackBot, Eaglizard, Evanreyes, Commander Keane bot, The monkeyhate, Timneu22, Bbq332, Dual Freq, Hgrosser, Dethme0w, Sergio.ballestrero, Wen D House, Radagast83, KazuiKier~enwiki, Xiutwel, Efilnickufesin, ArglebargleIV, BerndJantzen, Darktemplar, JorisvS, NNemec, Mets501, JdH, Yuide, Ossipewsk, Tmangray, W0lfie, Bhartsell, CRGreathouse, Nysin, Geremia, Van helsing, Edgerck, Holyfool, Adailton, Thijs!bot, Mbell, Headbomb, BehnamFarid, D.H, AntiVandalBot, WinBot, Benjaburns, Chgros, Rico402, Steelpillow, JAnDbot, Roman à clef, Petecarney, FK65, Yill577, Email4mobile, Boffob, InvertRect, Taborgate, Pagw, R'n'B, StorminMormon, Speed8ump, Lantonov, Jcwf, Ken g6, Omniver, Tkgd2007, VolkovBot, JabberWalkie, TallNapoleon, Holme053, DancingMan, Davehi1, Anonymous Dissident, Cloudswrest, Bearian, Arcfrk, SieBot, Macwillson, Radon210, Shrommer, Dhatfield, Lightmouse, Promodulus, JustBeCool, Anchor Link Bot, JKBlock, AussieScribe, SallyForth123, ClueBot, Bobathon71, Champion sound remix, Balashpersia~enwiki, Mild Bill Hiccup, VortexZero, Mgherm, Mumiemonstret, Kamilton, Brews ohare, M carteron, Jsondow, Bjdehut, Venkat11, Wnt, DumZiBoT, Grantscharoff, Booger1111, Lingjun.wang, A.Cython, V-squared, Gravitophoton, EjsBot, Download, CarsracBot, AnnaFrance, Cesiumfrog, David0811, Serespdi, Legobot, Luckas-bot, Yobot, Tohd8BohaithuGh1, TaBOT-zerem, Julia W, Aldebaran66, Wireader, Lildyson314, N1RK4UDSK714, AnomieBOT, Materialscientist, Citation bot, Quebec99, Rygbi7777, Celestinac, Omnipaedista, Blinduprising1, FrescoBot, Paine Ellsworth, Sławomir Biały, Alxeedo, Coder0xff2, Waxagass, N4tur4le, Jonesey95, Tom.Reding, Casimir9999, Philémon Cyclone, Canuck100, Ti-30X, Marie Poise, RjwilmsiBot, EmausBot, 𝕯𝖆𝖗𝖐, Quantanew, Tommy2010, Ornithikos, ZéroBot, Makecat, RockMagnetist, ClueBot NG, Siaraman, Widr, Bibcode Bot, BG19bot, Guy vandegrift, Mark Arsten, Ricky90007, BattyBot, The Illusive Man, ChrisGualtieri, Dexbot, Denysbondar, Andyhowlett, Nixon057, Leprof 7272, Anatolii j, Jwoodward48wiki, SaigonBlue, Zlelik2000, Soham, Anrnusna, Monkbot, Prisencolin, Yves Dubugnon, Heuh0, History is ma bae123, Casimir7, Lolgirl134, Aetherdisplacement and Anonymous: 235

- **Earnshaw's theorem** *Source:* https://en.wikipedia.org/wiki/Earnshaw'{}s_theorem?oldid=677400422 *Contributors:* The Anome, William Avery, Michael Hardy, Charles Matthews, Omegatron, Ojigiri~enwiki, Giftlite, Wolfkeeper, Syntaxers, MFNickster, Hellisp, TheBlueWizard, Rich Farmbrough, Bender235, Eruantalon, Jérôme, Wtshymanski, Oleg Alexandrov, Woohookitty, Ruud Koot, Mathbot, Ahpook, Splash, Netscott, Ridiculous fish, Newagelink, Light current, Besselfunctions, Roland Longbow, Stepa, Thumperward, Henning Makholm, Mgiganteus1, SMasters, Chetvorno, WinBot, Compbiowes, Deans-nl, HEL, VolkovBot, Abaglos, Yilloslime, Anton Gutsunaev, AlleborgoBot, SieBot, Wing gundam, Finmar36, GregVolk, Addbot, Aboctok, AnomieBOT, 4e to 4e, GliderMaven, FrescoBot, Kyteto, Maxus96, RedBot, Dinamikbot, J-p-fm, EmausBot, Wikipelli, ZéroBot, Akerans, Ebrambot, Teapeat, ClueBot NG, Bibcode Bot, YFdyh-bot, Andyhowlett, 20M030810, Sparx412 and Anonymous: 39

- **Electrogravitics** *Source:* https://en.wikipedia.org/wiki/Electrogravitics?oldid=699978472 *Contributors:* Reddi, AlainV, Apotheon, Pjacobi, MJT1331, I9Q79oL78KiL0QTFHgyc, Zetawoof, Pharos, Confusedmiked, Alansohn, RJFJR, Bobrayner, Linas, Uncle G, BillC, GregorB, Qwertyus, Rjwilmsi, Isotope23, Bgwhite, Hillman, RussBot, Bhny, Cryptic, DouglasHeld, JulesH, Sfnhltb, Kortoso, Petri Krohn, SmackBot, Oli Filth, Revelations, Nima Baghaei, DMacks, Bcasterline, JzG, Danielsdoug, Perfectblue97, Ocatecir, Dicklyon, Meco, Ambuj.Saxena, Norm mit, Cbrown1023, Satori Son, Second Quantization, Mentifisto, Sophie means wisdom, R'n'B, Fountains of Bryn Mawr, Juliancolton, Puddytang, GLPeterson, Spinningspark, Rep07, Biscuittin, Keilana, Nike787, Lukefortune, Timirdatta, Rhododendrites, Addbot, Simonm223, Download, Tassedethe, Verbal, Yobot, Aldebaran66, AnomieBOT, Autogenesis, Citation bot, LilHelpa, Turnvater Jahn, FrescoBot, Anothroskon, The Strategist, Rcsprinter123, Sbmeirow, JulioMarco, AerobicFox, Snotbot, Frietjes, Helpful Pixie Bot, Wildman741990, United States Man, Dexbot, Linoavac, Ehammer2, Dannis243 and Anonymous: 82

- **Electrostatic levitation** *Source:* https://en.wikipedia.org/wiki/Electrostatic_levitation?oldid=663329458 *Contributors:* Bryan Derksen, The Anome, Lommer, Charles Matthews, Dcoetzee, Reddi, Omegatron, Alansohn, BRW, Axeman89, Mandarax, Fragglet, Netscott, Spike Wilbury, Cosmotron, SmackBot, Bluebot, PointyOintment, Ryulong, Iridescent, CmdrObot, Alaibot, Dlabtot, Polyamorph, Aboctok, Termar, Yobot, Paine Ellsworth, Pinethicket, Edderso, Stefan.K., Jesse V., Andrew Koza, Bomazi, ClueBot NG, GimmickNG and Anonymous: 16

- **Flux pumping** *Source:* https://en.wikipedia.org/wiki/Flux_pumping?oldid=684407959 *Contributors:* Orangemike, Ukexpat, Jnestorius, Gene Nygaard, Rjwilmsi, Kri, Bgwhite, SmackBot, Ohconfucius, Makyen, Chetvorno, Myasuda, Mblumber, A876, Davewho2, SchreiberBike, KitemanSA, Yobot, AnomieBOT, Citation bot, NOrbeck, J04n, Nerdseeksblonde, FrescoBot, Citation bot 1, Tom.Reding, Marcin Sablik, RjwilmsiBot, Set theorist, Bibcode Bot, Monkbot, SkateTier and Anonymous: 9

- **Viktor Grebennikov** *Source:* https://en.wikipedia.org/wiki/Viktor_Grebennikov?oldid=693494355 *Contributors:* Big iron, Reddi, DragonflySixtyseven, El stiko, Sfacets, GregorB, Czar, Jaraalbe, Conscious, Alex Bakharev, Howcheng, Chris the speller, Bluebot, JzG, Ryulong, Cydebot, Thijs!bot, Barticus88, Dsp13, Waacstats, Macassar, EoGuy, PixelBot, Addbot, Atethnekos, Lightbot, OlEnglish, Aldebaran66, GenQuest, Tom.Reding, Niente0, RjwilmsiBot, Bobtheblueberry, Pekpekpek, Widr, MrBill3, VIAFbot, Monkbot, KasparBot and Anonymous: 4

- **Ionocraft** *Source:* https://en.wikipedia.org/wiki/Ionocraft?oldid=698270639 *Contributors:* Derek Ross, Thesteve, Mdupont, Glenn, Samw, Reddi, David Latapie, Omegatron, Finlay McWalter, Robbot, Centrx, JetJon, AlistairMcMillan, Geni, Scott Burley, Rich Farmbrough, Luvcraft, Pjacobi, Kbh3rd, Shanes, Femto, Keron Cyst, Hooperbloob, Mgaved, Mr Adequate, BRW, Gene Nygaard, Blaze Labs Research, Timharwoodx, GalFisk, Rjwilmsi, Wikibofh, Bgwhite, Bhny, Splash, Netscott, Gaius Cornelius, Aaron Schulz, Ilmaisin, Closedmouth, Peter, Curpsbot-unicodify, Nekura, Anthony717, SmackBot, Stux, Teemu Ruskeepää, Chris the speller, Joefaust, Beerathon, MovGP0, Trekphiler, Fuhghettaboutit, Nishkid64, Tktktk, Hikui87~enwiki, Coffee Atoms, A876, Alanbly, Omicronpersei8, Thijs!bot, Bobblehead, Ileresolu, Akradecki, Tim Shuba, CosineKitty, Magioladitis, Faizhaider, Alan2here, Ubigcow, Dispenser, Fountains of Bryn Mawr, Equazcion, SirBob42, Inwind, James Kidd, VolkovBot, Robinson weijman, Eddie mars, Thunderbird2, Agwholland, Biscuittin, Flyer22 Reborn, Tiptoety, MichaelJPierce, Binksternet, Feyre, The Thing That Should Not Be, Kendo70133, Alchemist Jack, Shpakovich, Addbot, Gregz08, Lightbot, Anand.Hegde, AnomieBOT, Censorship Workaround, FrescoBot, RandomDSdevel, Joshuachohan, Russell Anderson, RaptureBot, BrokenAnchorBot, Shrikanthv, Harulover, QuantumSquirrel, ClueBot NG, Jarmicols, The High Fin Sock Whale, Jdperkins, Helpful Pixie Bot, BG19bot, Meatsgains, Dexbot, FoCuSandLeArN, Randykitty, CameronCoe, Mike Mounier, Monkbot, Eman235, NathanSimkiss and Anonymous: 123

- **Optical levitation** *Source:* https://en.wikipedia.org/wiki/Optical_levitation?oldid=680625603 *Contributors:* Arpingstone, Samw, BRW, Chobot, Netscott, PointyOintment, Jeff Wheeler, Headbomb, WingkeeLEE, CohesionBot, Addbot, Shawn Worthington Laser Plasma, Monkbot and Anonymous: 5

- **Squeezed vacuum** *Source:* https://en.wikipedia.org/wiki/Squeezed_vacuum?oldid=663626240 *Contributors:* Rpyle731, Quantanew and Wgolf

- **Superdiamagnetism** *Source:* https://en.wikipedia.org/wiki/Superdiamagnetism?oldid=589602981 *Contributors:* Michael Hardy, Phys, Xanzzibar, Kongsvold, Beland, Gene Nygaard, PoccilScript, Chobot, Netscott, SmackBot, E. Sn0 =31337=, Saippuakauppias, Siddharth srinivasan, Andrew Hampe, Zero sharp, Ale jrb, Myasuda, Kareemjee, Rifleman 82, Xenocide321, Tikiwont, VolkovBot, YonaBot, Crazz bug 5, Addbot, Lightbot, HRoestBot, DARTH SIDIOUS 2, Circuitboardsushi, Asdf995, Tls60, Teapeat, Rememberway, ArthurDent006.5 and Anonymous: 9

## 13.4.2  Images

- **File:A_maglev_train_coming_out,_Pudong_International_Airport,_Shanghai.jpg** *Source:* https://upload.wikimedia.org/wikipedia/commons/ d/d1/A_maglev_train_coming_out%2C_Pudong_International_Airport%2C_Shanghai.jpg *License:* Public domain *Contributors:* Originally from en.wikipedia; description page is (was) here *Original artist:* User Alex Needham (own photography) on en.wikipedia

- **File:Ambox_important.svg** *Source:* https://upload.wikimedia.org/wikipedia/commons/b/b4/Ambox_important.svg *License:* Public domain *Contributors:* Own work, based off of Image:Ambox scales.svg *Original artist:* Dsmurat (talk · contribs)

- **File:Casimir_plates.svg** *Source:* https://upload.wikimedia.org/wikipedia/commons/4/44/Casimir_plates.svg *License:* CC BY-SA 3.0 *Contributors:* Own work *Original artist:* Emok

- **File:Commons-logo.svg** *Source:* https://upload.wikimedia.org/wikipedia/en/4/4a/Commons-logo.svg *License:* ? *Contributors:* ? *Original artist:* ?

- **File:E_J_Power_Law.JPG** *Source:* https://upload.wikimedia.org/wikipedia/commons/7/71/E_J_Power_Law.JPG *License:* GFDL *Contributors:* E_J Law Simulations *Original artist:* Mark Ainslie

- **File:EfektMeisnera.svg** *Source:* https://upload.wikimedia.org/wikipedia/commons/b/b5/EfektMeisnera.svg *License:* Public domain *Contributors:* Inspiration: Image:EXPULSION.png *Original artist:* Piotr Jaworski, <a href='//pl.wikipedia.org/wiki/Wikipedysta:Piom' class='extiw' title='pl:Wikipedysta:Piom'>*PioM*</a> EN DE PL; POLAND/Poznań

- **File:Electrostatic_Levitation_of_a_Titanium-Zirconium-Nickel_Alloy.jpg** *Source:* https://upload.wikimedia.org/wikipedia/commons/ 7/79/Electrostatic_Levitation_of_a_Titanium-Zirconium-Nickel_Alloy.jpg *License:* Public domain *Contributors:* nasaimages.org (alternate) *Original artist:* NASA/MSFC/Emmett Given

### 13.4.3   Content license